Hands On History
A Resource for Teaching Mathematics

© *2007 by*
The Mathematical Association of America (Incorporated)

Library of Congress Catalog Card Number 2007937009

ISBN 978-0-88385-182-1

Printed in the United States of America

Current Printing (last digit):
10 9 8 7 6 5 4 3 2 1

Hands On History
A Resource for Teaching Mathematics

Edited by

Amy Shell-Gellasch
Pacific Lutheran University

Published and Distributed by
The Mathematical Association of America

The MAA Notes Series, started in 1982, addresses a broad range of topics and themes of interest to all who are involved with undergraduate mathematics. The volumes in this series are readable, informative, and useful, and help the mathematical community keep up with developments of importance to mathematics.

Council on Publications
James Daniel, *Chair*

Notes Editorial Board
Stephen B Maurer, *Editor*
Paul E. Fishback, *Associate Editor*

Michael C. Axtell   Rosalie Dance   William E. Fenton
Donna L. Flint   Michael K. May   Judith A. Palagallo
Mark Parker   Susan F. Pustejovsky   Sharon Cutler Ross
David J. Sprows   Andrius Tamulis

MAA Notes

14. Mathematical Writing, by *Donald E. Knuth, Tracy Larrabee, and Paul M. Roberts*.
16. Using Writing to Teach Mathematics, *Andrew Sterrett*, Editor.
17. Priming the Calculus Pump: Innovations and Resources, Committee on Calculus Reform and the First Two Years, a subcommittee of the Committee on the Undergraduate Program in Mathematics, *Thomas W. Tucker*, Editor.
18. Models for Undergraduate Research in Mathematics, *Lester Senechal*, Editor.
19. Visualization in Teaching and Learning Mathematics, Committee on Computers in Mathematics Education, *Steve Cunningham and Walter S. Zimmermann*, Editors.
20. The Laboratory Approach to Teaching Calculus, *L. Carl Leinbach et al.*, Editors.
21. Perspectives on Contemporary Statistics, *David C. Hoaglin and David S. Moore*, Editors.
22. Heeding the Call for Change: Suggestions for Curricular Action, *Lynn A. Steen*, Editor.
24. Symbolic Computation in Undergraduate Mathematics Education, *Zaven A. Karian*, Editor.
25. The Concept of Function: Aspects of Epistemology and Pedagogy, *Guershon Harel and Ed Dubinsky*, Editors.
26. Statistics for the Twenty-First Century, *Florence and Sheldon Gordon*, Editors.
27. Resources for Calculus Collection, Volume 1: Learning by Discovery: A Lab Manual for Calculus, *Anita E. Solow*, Editor.
28. Resources for Calculus Collection, Volume 2: Calculus Problems for a New Century, *Robert Fraga*, Editor.
29. Resources for Calculus Collection, Volume 3: Applications of Calculus, *Philip Straffin*, Editor.
30. Resources for Calculus Collection, Volume 4: Problems for Student Investigation, *Michael B. Jackson and John R. Ramsay*, Editors.
31. Resources for Calculus Collection, Volume 5: Readings for Calculus, *Underwood Dudley*, Editor.
32. Essays in Humanistic Mathematics, *Alvin White*, Editor.
33. Research Issues in Undergraduate Mathematics Learning: Preliminary Analyses and Results, *James J. Kaput and Ed Dubinsky*, Editors.
34. In Eves' Circles, *Joby Milo Anthony*, Editor.
35. You're the Professor, What Next? Ideas and Resources for Preparing College Teachers, The Committee on Preparation for College Teaching, *Bettye Anne Case*, Editor.
36. Preparing for a New Calculus: Conference Proceedings, *Anita E. Solow*, Editor.
37. A Practical Guide to Cooperative Learning in Collegiate Mathematics, *Nancy L. Hagelgans, Barbara E. Reynolds, SDS, Keith Schwingendorf, Draga Vidakovic, Ed Dubinsky, Mazen Shahin, G. Joseph Wimbish, Jr.*
38. Models That Work: Case Studies in Effective Undergraduate Mathematics Programs, *Alan C. Tucker*, Editor.
39. Calculus: The Dynamics of Change, CUPM Subcommittee on Calculus Reform and the First Two Years, *A. Wayne Roberts*, Editor.
40. Vita Mathematica: Historical Research and Integration with Teaching, *Ronald Calinger*, Editor.
41. Geometry Turned On: Dynamic Software in Learning, Teaching, and Research, *James R. King and Doris Schattschneider*, Editors.

42. Resources for Teaching Linear Algebra, *David Carlson, Charles R. Johnson, David C. Lay, A. Duane Porter, Ann E. Watkins, William Watkins,* Editors.
43. Student Assessment in Calculus: A Report of the NSF Working Group on Assessment in Calculus, *Alan Schoenfeld,* Editor.
44. Readings in Cooperative Learning for Undergraduate Mathematics, *Ed Dubinsky, David Mathews, and Barbara E. Reynolds,* Editors.
45. Confronting the Core Curriculum: Considering Change in the Undergraduate Mathematics Major, *John A. Dossey,* Editor.
46. Women in Mathematics: Scaling the Heights, *Deborah Nolan,* Editor.
47. Exemplary Programs in Introductory College Mathematics: Innovative Programs Using Technology, *Susan Lenker,* Editor.
48. Writing in the Teaching and Learning of Mathematics, *John Meier and Thomas Rishel.*
49. Assessment Practices in Undergraduate Mathematics, *Bonnie Gold,* Editor.
50. Revolutions in Differential Equations: Exploring ODEs with Modern Technology, *Michael J. Kallaher,* Editor.
51. Using History to Teach Mathematics: An International Perspective, *Victor J. Katz,* Editor.
52. Teaching Statistics: Resources for Undergraduate Instructors, *Thomas L. Moore,* Editor.
53. Geometry at Work: Papers in Applied Geometry, *Catherine A. Gorini,* Editor.
54. Teaching First: A Guide for New Mathematicians, *Thomas W. Rishel.*
55. Cooperative Learning in Undergraduate Mathematics: Issues That Matter and Strategies That Work, *Elizabeth C. Rogers, Barbara E. Reynolds, Neil A. Davidson, and Anthony D. Thomas,* Editors.
56. Changing Calculus: A Report on Evaluation Efforts and National Impact from 1988 to 1998, *Susan L. Ganter.*
57. Learning to Teach and Teaching to Learn Mathematics: Resources for Professional Development, *Matthew Delong and Dale Winter.*
58. Fractals, Graphics, and Mathematics Education, *Benoit Mandelbrot and Michael Frame,* Editors.
59. Linear Algebra Gems: Assets for Undergraduate Mathematics, *David Carlson, Charles R. Johnson, David C. Lay, and A. Duane Porter,* Editors.
60. Innovations in Teaching Abstract Algebra, *Allen C. Hibbard and Ellen J. Maycock,* Editors.
61. Changing Core Mathematics, *Chris Arney and Donald Small,* Editors.
62. Achieving Quantitative Literacy: An Urgent Challenge for Higher Education, *Lynn Arthur Steen.*
64. Leading the Mathematical Sciences Department: A Resource for Chairs, *Tina H. Straley, Marcia P. Sward, and Jon W. Scott,* Editors.
65. Innovations in Teaching Statistics, *Joan B. Garfield,* Editor.
66. Mathematics in Service to the Community: Concepts and models for service-learning in the mathematical sciences, *Charles R. Hadlock,* Editor.
67. Innovative Approaches to Undergraduate Mathematics Courses Beyond Calculus, *Richard J. Maher,* Editor.
68. From Calculus to Computers: Using the last 200 years of mathematics history in the classroom, *Amy Shell-Gellasch and Dick Jardine,* Editors.
69. A Fresh Start for Collegiate Mathematics: Rethinking the Courses below Calculus, *Nancy Baxter Hastings,* Editor.
70. Current Practices in Quantitative Literacy, *Rick Gillman,* Editor.
71. War Stories from Applied Math: Undergraduate Consultancy Projects, *Robert Fraga,* Editor.
72. Hands On History: A Resource for Teaching Mathematics, *Amy Shell-Gellasch,* Editor.

MAA Service Center
P.O. Box 91112
Washington, DC 20090-1112
1-800-331-1MAA   FAX: 1-301-206-9789

Preface

While teaching at the United States Military Academy (USMA) at West Point, New York from 2000 to 2003, I had the immense good fortune to work with Fred Rickey on the history of the Olivier String Models owned by the USMA Department of Mathematical Sciences. Working with and researching these models[1] became a joy for me. One question led to another, and I became acquainted with all aspects of the models.

To understand the models and present the mathematical concepts they depict to students, I had to learn the mathematics involved in both current and historical terms. I also had to learn about those who developed the mathematics as well as those who developed and constructed the models. One of the aspects of this work that I found the most intriguing was tracking the history of the actual models: who constructed them, who bought them, where have they been since their construction, etc. In many cases, this was the most difficult part of the research. Along the way, I learned a lot of history about the Academy itself and the mathematics department while researching the models through the USMA archives. Fred and I had great fun searching the department for a few missing models that the archives said we should have. We did find one, only a picture of another, and alas, never did find a third. But we also found many other old and forgotten teaching aides, such as a chalk board globe, which we enjoyed bringing back into use in the classroom. Handling the models and bringing them back to working order was very enjoyable, while creating public displays for the models and presenting them to students and faculty allowed me to share these experiences with others.[2] Finally, I was inspired by the simple beauty of these mathematical models and now see the beauty in many other mathematical models and objects. Some mathematicians, like my colleague Bill Acheson, have been inspired to construct models themselves, while artists have been inspired by the clean lines of the string models to use those elements in string sculptures.

All this led me to organize a MAA short course with Glen Van Brummelen at the 2004 Joint Mathematics Meetings in Phoenix, entitled The History of Mathematical Technologies: Exploring the Material Culture of Mathematics. I believe this short course was a great success, and led to this project. Many of the speakers who so graciously shared their expertise during those two days have contributed to this volume. I have added to their ranks other historians, researchers and professionals with an interest in the physical embodiment of mathematics and a wish to use that to enhance the teaching of mathematics. Several of these individuals were in the audience of the short course.

All of us involved in this volume have a love of the physical manifestation of mathematics, a love of teaching, and a desire to bring the two together. I would like to thank all the contributors for sharing their knowledge and love of mathematical models. Finally, I would like to thank the MAA staff for their energy and dedication in creating the final product, and the readers for their time, care and diligence in giving helpful, insightful and honest feedback.

Amy Shell-Gellasch
October 2006

---

[1] USMA has twenty-four models constructed in the 1850s by French mathematician Theodore Olivier to depict concepts from descriptive geometry—see the sixth chapter in this volume.

[2] See the chapter by Kidwell and Ackerberg-Hastings on creating displays.

Introduction

With the advent of computers and classroom projectors, educators have a world of resources at their fingertips. However, when we let our fingers do the walking in the virtual world, we and our students miss the physical and creative aspects of learning. Numerous studies have shown that doing (as opposed to simply listening or reading) is the best way to learn.

Educators have always employed physical models to aid in instruction. During much of the nineteenth century and the twentieth up to the advent of classroom computers, makers of mathematical devices for the classroom abounded. See *Multi-Sensory Aids in Teaching Mathematics* (1966) of The National Council of Teachers of Mathematics, and reprinted by the American Mathematical Society for a look at the types and diversity of mathematical models and apparatus used in classrooms during the first half of the twentieth century.

Historically, scientists and mathematicians have constructed models to help them explore the physical world, be it in engineering, chemistry, physics, astronomy, or the realm of pure mathematics. In addition, the daily practice and use of mathematics was tied to the use of devices such as counting boards and the abacus, and in science with the use of astrolabes, quadrant finders, and the like. The physical items used both in the practice of mathematics and its teaching, are an important part of our mathematical heritage. I believe exploring these items adds a level of understanding that is not available with computer simulations. Students can construct and manipulate a physical device to immediately explore their own questions and follow their own creativity in a much more personal and tactile way than simply typing commands into a keyboard or moving a mouse.

This volume is meant to be a resource guide to help school and collegiate educators incorporate history into their classrooms in the form of historical devices. In our increasingly electronic society, students are losing "touch" with the physical aspects of mathematics and science. By bringing actual historical items into the classroom or by constructing replicas, students will be engaged in their learning in a more concrete manner.

The papers in this volume represent a wide variety of topics and levels of mathematics as well as levels of difficulty of construction. They include short, one-day projects for the high school or junior high classroom as well as longer projects to be constructed by motivated college students. All of the ideas presented here can also be used with student mathematics clubs or as individual or small group projects. I hope that the selection will encourage you to get back in *touch* with the physical heritage of mathematics and share that heritage with your students.

Contents

Preface vii

Introduction ix

Learning from the Medieval Master Masons: A Geometric Journey through the Labyrinth 1
   Hugh McCague

Dem Bones Ain't Dead: Napier's Bones in the Classroom 17
   Joanne Peeples

The Towers of Hanoi 29
   Amy Shell-Gellasch

Rectangular Protractors and the Mathematics Classroom 35
   Amy Ackerberg-Hastings

Was Pythagoras Chinese? 41
   David E. Zitarelli

Geometric String Models of Descriptive Geometry 49
   Amy Shell-Gellasch and Bill Acheson

The French Curve 63
   Brian J. Lunday

Area Without Integration: Make Your Own Planimeter 71
   Robert L. Foote and Ed Sandifer

Historical Mechanisms for Drawing Curves 89
   Daina Taimina

Learning from the Roman Land Surveyors: A Mathematical Field Exercise 105
   Hugh McCague

Equating the Sun: Geometry, Models, and Practical Computing in Greek Astronomy 115
   James Evans

Sundials: An Introduction to Their History, Design, and Construction 125
   J. L. Berggren

Why is a Square Square and a Cube Cubical? 139
   Amy Shell-Gellasch

The Cycloid Pendulum Clock of Christiaan Huygens 145
   Katherine Inouye Lau and Kim Plofker

Build a Brachistochrone and Captivate Your Class 153
   V. Frederick Rickey

Exhibiting Mathematical Objects: Making Sense of your Department's Material Culture 163
   Peggy Aldrich Kidwell and Amy Ackerberg-Hastings

About the Authors 175

Learning from the Medieval Master Masons: A Geometric Journey through the Labyrinth

Hugh McCague
York University

Introduction

In recent years, there has been a resurgence in understanding, constructing, and walking mazes and labyrinths, many originating from the Middle Ages. For the student of mathematics, these novel geometric objects can be an experience in the power, beauty and utility of mathematics.

These qualities of mathematics were well appreciated by medieval master masons, the designers and builders of the great pavement labyrinths in some of the medieval churches and cathedrals. Master masons worked their way up the ranks of stonemasonry starting with demanding years of apprenticeship. There were many tools for stone cutting and sculpting to master. The key and emblematic tools for stonemasonry were large compass-dividers, set square [Figure 1], and measuring rod. These tools underscore the master masons' skill in practical geometry, not in theoretical mathematics involving axioms, theorems and proofs [1]. Common mathematical and geometrical motifs of the master masons [2] were:

   a) a square rotated a half of a right angle about its center,

   b) the square's side to its diagonal (in modern terms, unknown to the medieval mason, $1:\sqrt{2}$),

   c) the equilateral triangle's half-side to its altitude (in modern terms $1:\sqrt{3}$),

   d) the regular pentagon's side to its diagonal (in modern terms the 'golden section' and $\sqrt{5}-1:2$), and

   e) simple whole number ratios, e.g., 1 : 1, 1 : 2, and 3 : 4.

Designs were translated to full-scale templates, made of cloth and wood, for the cutting of limestone, sandstone, and marble [3]. Mathematics reigned throughout the process: practical geometry with those motifs allowed a repertoire of designs to be memorized and varied, and easily communicated to mason assistants. Indeed, in the earliest extant codified rules of this craft, dating from c. 1390–1400, stonemasonry is referred to as the "Art of Geometry" [4].

Master masons designed, guided and supervised the stonework of wide-ranging projects from well-to-do houses to castles, from town wharfs to fortified city walls. A major part of their work was churches and cathedrals including stone pavement labyrinths.

The central formative story of labyrinths, important to these medieval pavement labyrinths, was the legend of Daedalus designing and building the ancient Cretan labyrinth that Theseus traversed in order to slay the Minotaur at the center [5]. The legend, as with records of later labyrinths, does not relate the historical mathematics applied in the construction. As a result, we must infer those methods from the extant labyrinths themselves and our knowledge of the working methods and mathematical knowledge of earlier craftspersons such as the medieval masons.

During the construction of some of the French Cathedrals of the 13th century, masterworks of labyrinths were laid by masons in the pavements of the naves [Figure 2]. Only a moderate amount of historical documentation survives regarding the medieval masons during the 13th century and earlier, so much of our knowledge of the work by masons on labyrinths needs to be inferred from the buildings and labyrinths themselves. For example, the well-known circular pavement labyrinth at Chartres Cathedral (1194–c.1221) dates from c.1215–c.1221. The path's border is made of blue-black marble, the path itself is made of 276 white limestone slabs, and the outer diameter is 42.16′ (12.85m) [6]. Amiens Cathedral (1220–1288) had an octagonal pavement labyrinth laid by Master Renaud de Cormont in 1288, but it was removed during 1827–1828. An exact replica was laid back in place in 1894 [Figure 2]. White stones mark the border, black stones form the path, and the structure spans 39.83′ (12.14m) [7]. The octagonal pavement labyrinth at Reims Cathedral (1211–c.1290) dated from c. 1290, but was destroyed in 1779. We are fortunate to have some drawings that recorded this labyrinth's design which included octagonal 'pods' in the four corners [Figure 4]. This labyrinth had a white stone border and a black marble path. The width of the structure in its entirety was approximately 34′ (10.2m). Of special note to us are the outer four octagonal 'pods' which commemorated and depicted the Cathedral's successive master masons with their mathematical instruments [8].

Figure 1. King Offa and a master mason, with large set square and compass in hand, directing stone-masons at St. Albans Cathedral, from Matthew Paris, *Life of St. Alban,* c. 1245–52, Dublin, Trinity College Library, MS 177, fol.59v. Reproduced courtesy of The Board of Trinity College Dublin.

What is a labyrinth or maze?

A maze is a collection of connected paths that lead you to dead ends, a center or rest areas, and exits. The term 'labyrinth' here will be used to mean a maze with one path, a unicursal maze, that winds about itself and leads to a center without alternative routes and additional dead ends. This definition of labyrinth, preferred by many labyrinth practitioners, is not standard and, in general, the terms maze and labyrinth can be used interchangeably. A labyrinth or maze can have vertical walls to make the structure 3-dimensional. The structures we will work with are floor or pavement labyrinths, unicursal mazes, in 2 dimensions.

We will follow step-by-step geometric instructions to recreate the design of the large pavement labyrinth at the great Gothic Cathedral at Amiens in northern France. Rather than working with stone as in the original, I will suggest some simpler inexpensive materials. The original pavement labyrinth at Amiens Cathedral was, as noted already, 39.83′ (12.14m) across. I have scaled the size down to 14′ because this is often a suitable size for fitting the labyrinth into a classroom. Additionally, as we will see, 14 has some helpful mathematical properties for laying out the design. You can scale up or down from 14′ to make the labyrinth best suited for your rooms and usages. The path of a scaled-down labyrinth will of course be relatively narrow. As you walk such labyrinths, you need to center your attention on the path, and do not be concerned that your feet may be protruding over borders of the narrow path. You can also simply scale and construct the labyrinth with pencil on letter-size paper and trace the path with your fingers.

Materials and geometric tools for making labyrinths

The design can be laid out on large sheets of canvas (sewn together) or other material on the floor of the classroom or gymnasium or on the playground rather than on a stone floor of a cathedral. Large sheets of canvas can be purchased at canvas specialty companies, larger art supply stores, or at some home supply

Figure 2. View of the interior of Amiens Cathedral looking to the east. The pavement labyrinth, spanning the fourth and fifth bays from the west portal, is in the lower part of the picture. The path of the labyrinth consists of black stones with a border of white stones. Photograph by Stephen Murray.

stores. Canvas is fairly strong and lasts well, but it is good to get a thickness greater than that used in small canvases that are stretched for paintings. A good quality double-sided carpet tape should be used to hold down the canvas stretched out smooth and tight on the floor. Otherwise the canvas will be loose with small bumps that frustrate laying out an accurate design. For the borders or sides of the labyrinth's path, the better quality black electrical tape works well on an off-white canvas, both in terms of visual contrast and in providing a good adhesive contact for longer term use. When laying the electrical tape down you may wish to stretch it to get a good straight line, making sure you take the tension off the tape when you finally press it on the canvas. If the tape is even slightly stretched when applied to the surface you will find that, when you come back the next day, the tape has contracted and has pulled up from the canvas. The trimming of the tape can be done with a small pair of sharp scissors or, even better, an 'exacto' (X-Acto) knife with the tape fully in place. Be sure to use light pressure, so only the tape is cut and not the canvas underneath. The tape may be much narrower than the borders of the original stone pavement labyrinth such as at Amiens, even taking into account the scaled-down size of the canvas labyrinth. The essential issue is that the border be well marked in defining the path, and black electrical tape can do this job well.

Alternatively, for a short-term and non-portable labyrinth, you may wish to lay out the structure with chalk on pavement. Side-walk chalk which washes off with water or rain may be preferable, if you need the labyrinth cleared off soon afterwards. Some people get so they can rapidly lay out a labyrinth with masking tape that can then be pulled up shortly afterwards from a gym floor or similar surface [9]. Of course, test the surface ahead of time with smaller pieces of masking tape to insure the surface is not damaged upon the removal of the tape.

If you are making the labyrinth for short-term use, you may wish to work at a pace that accepts moderate inaccuracies. The mathematical challenge is recognizing that some inaccuracies are acceptable while others will lead to difficulties. For this reason, even if you need to get the labyrinth 'up' in a few hours, check everything as you proceed. Do not expect to get a sizable masking-tape labyrinth laid out in a few hours until you have had a fair amount of practice.

We will use mathematical tools, some of which are essentially the same as those used by the medieval master masons. If you are laying out with chalk on a paved surface such as a school playground, you may wish to use a chalkline available at hardware stores. This device coats a stretched cord with chalk which when snapped down on the surface leaves a good quality straight line segment. This same basic procedure was used by the ancient Egyptians and Greeks [10]. It takes some practice to become proficient, but it soon becomes a rapid way to lay out straight line segments on larger surfaces. A chalkline is not suitable for canvas because of the smudges to the material. A longer-term option on a school playground, once the guiding lines are set, would be the use of an appropriate paint, but first obtain advice from a hardware or paint store.

In addition to the chalkline, there are other ways to draw longer straight line segments. A stretched string, without a chalk coating, provides a good straight line. There is no firm side to run your pencil against. Nevertheless, one can do a fair job running the pencil beside the string, but not pushing the string off course. Yard rules and meter sticks are very handy, but often too short for part of the construction. One can make longer wooden or metal rods with a rectangular, not circular, cross-section for accurate drawing. A set square is invaluable for drawing and checking right angles.

The medieval master masons had, as mentioned earlier, large compasses or dividers that stood about a yard high [Figure 1]. Such big compasses do not seem readily and inexpensively available today, though a person skilled in woodworking could make one. Smaller compasses can be used to check work done, such as constructed angles, but these instruments are not generally accurate enough for this large scale work. A swung string fixed at a pole can do quite well for making larger arcs. However, one does not have to use a compass or its equivalent for the particular labyrinth design that we are about to undertake.

Craftspersons normally remove evidence of their earlier layout work. If you wish to later remove the guiding pencil marks, lightly use a pencil with a hard lead so the lines can be largely erased off the canvas. You may wish to leave the pencil marks intact so people can see direct evidence of the underlying geometrical process applied in this educational project.

Working methods and skill development

I suggest you view this exercise as a gradual learning process, not only in theoretical mathematics, but also in the application of mathematics that can give us an increased appreciation of the skill level of the medieval and later craftspersons. Your first labyrinth project may not be done as quickly and as accurately as you would like. If you make more labyrinths, you can increasingly apply what you have learned and it will be easier to guide others who are assisting you. Any related skills you and your helpers already have in building, landscaping, woodworking, machine shop work, art and graphic design, dress-making and other 'hands-on' applications of mathematics will definitely come to the fore to assist you.

We will see what geometric methods work practically, and maintain a reasonable accuracy. For accuracy it is generally best to work from larger dimensions to smaller dimensions. Thus, we will start with a large outer 14' square, and work inward from points on the outer square to determine crucial diagonals and axes as illustrated in Steps 1 to 4. Errors or inaccuracies tend to diminish as we move inward, but tend to expand as one works outward. I suggest that you triple check everything at each step. This approach helps insure that an error can be immediately rectified, rather than later realizing that the construction is not working and that you now have to back-track like a detective through successive stages of your work that need to be redone. One way to do extra checks on your work is to apply more than one mathematical approach. For example, once we have constructed the major diagonals of the concentric octagons illustrated in Step 3, check with a large protractor, a 45°-90°-45° drafting triangle, or by compass constructions (a smaller compass may not be so accurate, but it will still be a partial check) that the angles formed at the square's center are in fact multiples of 45° or half a right angle.

With some ingenuity, a larger labyrinth could be made by one person. However, it is good to have at least two people with one acting as the 'master' craftsperson guiding the project. Five to six people can make the construction even easier for, say, a labyrinth 14' across as here. For even larger labyrinths, more people can be well utilized. Keeping those persons active and working accurately throughout takes practice. For a class, it may be best to have students take turns working on different stages of the project, or to make more than one labyrinth. In the process, there is a marked sense of mathematical accomplishment, as the group sees the labyrinth 'manifest' or take form before them.

The geometric construction steps

Step 1

Lay out a square 14' by 14'. First, draw a 14' side. Use the Pythagorean triplet of 3-4-5 to create and check the right angles. You can use the triplet 3'-4'-5'. However, I suggest the triplet 3 yards-4 yards-5 yards which is a little more involved to layout, but further magnifies any error in the right angle.

Cathedral pavement labyrinths are commonly oriented eastward, though not necessarily exactly due east. Thus, the entrance (to be constructed in Step 8) is in the west. The east side of the labyrinth will be pictured at the top, even though

today's convention is to place north at the top. During the medieval period it was common to place east at the top of a map or spatially-oriented diagram. For this reconstruction of a medieval labyrinth we will do likewise.

From one end of the 14′ side measure 3 yards along that side. Next, from the first end (a corner of the square to be), measure out 4 yards and keep the measuring tape in place. Now with a second measuring tape, measure from the second end of the 3-yard length over to the 4-yard mark on the first measuring tape. Adjust the two measuring tapes, while straight and taut, until the 4-yard mark on the first tape coincides with the 5-yard mark on the second tape. This point of intersection is then used to lay out the 2nd side of the square which needs to be extended to 14′ to determine the 3rd corner of the square. In a similar fashion, the 4th corner and the 2 remaining sides of the square can be determined.

This constructed square is the crucial 'foundation' for the rest of the construction, so take time to get it accurate (say within 1/2″ on all sides). Also, check the square's two diagonals in feet at $14\sqrt{2} \approx 19.7989... \approx 19′\ 9\ 1/2″$ (close to 20′). If the diagonals are very close to this value (say within 3/4″), the right angles are sufficiently achieved, and this rhombus can be considered a square as required. Conceptually, the construction of a square is simple, but the accurate laying out of a large square takes practice and patience. If the sides or the diagonals are not very close to equal, or the attempt at right angles are slightly off, it can be a good mathematical problem to figure out what went wrong and how to correct the situation.

Step 2

Conceptually, a regular octagon can arise from the 8 intersection points of a square rotated a half of a right angle, or 45°, about its center. Generally, one is not using a big square that can be picked up and rotated accurately, so an equivalent procedure is required. On the initial square, the intersection points measured out on a side starting from a corner or vertex will occur approximately

    4.1005′ and 9.8995′ away, or
    4′ 1 1/4″ and 9′ 10 3/4″ away, and
    close to 4′ and 10′ away.

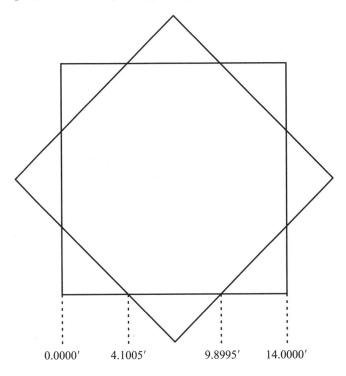

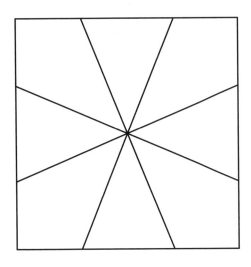

Using string, the medieval master mason could have swung the original square's 2 lower half-diagonals down (with centers at the 2 lower corners) to determine these same points on the square's side.

Step 3

Join the 8 intersection points to form 4 diagonals that pass through the center of the square. With some care and patience, one can get the 4 diagonals to actually intersect close together.

Learning from the Medieval Master Masons: A Geometric Journey through the Labyrinth

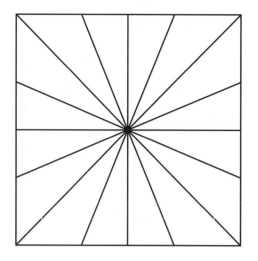

Step 4

Next draw the two diagonals of the square. Also, join, passing through the square's center, the midpoints of the square's sides to form the axes. If the layout has been careful, their intersection point should be very close to the square's center determined by the intersection of the other diagonals drawn in the previous step. If they are not close, go back over the earlier steps and check for their accuracy.

Step 5

We can now use all those line segments passing through the square's center to guide the octagonal geometry of the labyrinth and the location of all the turns in the path that we will be constructing. First, construct the inner regular octagon in the central region and allow enough room for the 11 other successively larger concentric regular octagons (as constructed in Step 6). A simple way to construct the regular octagons is to measure along one of the line segments that will become the major diagonals of these octagons. The length of half of one of these major diagonals AC is $7\sqrt{[1 + (\sqrt{2} - 1)^2]}' \approx 7.576745'$ $\approx 7'\ 6\ 15/16''$. We need to divide this length into space for the central small octagon and the 11 other octagons. For ease of construction, we use a path width (the distance between two successive octagons, illustrated in Step 6, as measured along AC) that is a whole number of inches plus possibly a half inch if needed. Also, we wish to keep the size of the inner octagon compared to the outer boundary of the labyrinth in similar proportion to the corresponding elements of the labyrinth at Amiens Cathedral. A solution that allows a reasonably sized central octagon and path width is: $7'\ 6\ 15/16'' = 1'\ 7\ 7/16'' + 11 \times 6\ 1/2''$.

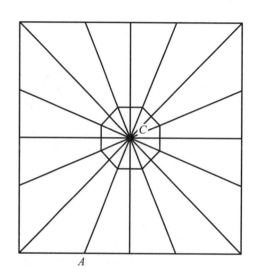

Path widths of 6" or 7" make the central octagon too big or too small, respectively. (You may find it helpful to look at the diagram for Step 6 to see the path width repeated 11 times, with widths of 6 1/2" as measured along AC, that will be formed by the 12 octagons that are about to be constructed.) Therefore, measure out from the square's center along AC 1' 7 7/16" and do similarly on the other 7 major half-diagonals for the octagons. (In other words, the length of the major half-diagonal of the first octagon is 1' 7 7/16".) These end-points can then be connected by line segments to form the first octagon and its vertices. Note the path width along AC is 6 1/2" = 6.5" (call it w_1 as illustrated in Step 8).

Step 6

From the first octagon, measure out along all 8 of the major half-diagonals (e.g. AC) of the octagons the first 11 multiples of 6.5" (an arithmetic sequence) to get the position of the suc-

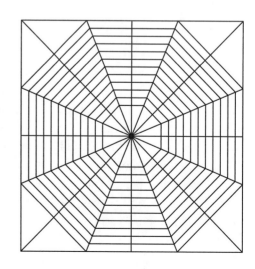

cessive path widths: 6.5″, 13″, 19.5″, …, 65″, 71.5″. With care, the last length of 71.5″ should land on or very close to the square's perimeter. Now connect these points with line segments to form the next 11 octagons to obtain a total of 12 concentric regular octagons. The basic framework for the labyrinth is now in place.

One could also measure the successive path widths from the square's perimeter to the inner octagon. In this case, working inward or outward does not make a difference in accuracy.

Step 7

Now, we need to put in the appropriate turns and openings in order to obtain the unicursal path that traverses the entire space of the labyrinth. In the diagram at right, the major diagonals of the octagon and square have been removed to simplify the explanatory picture. In the actual layout on canvas or other material, these lines in light pencil markings can be left or erased.

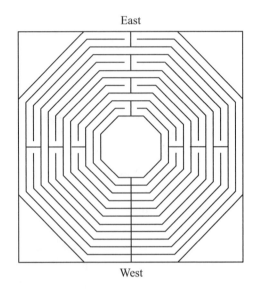

The path width w_2 (illustrated in Step 8), measured not along the major half-diagonals of the outer octagon (e.g., AC in Step 5 and 6), as in the case of $w_1 = 6.5$", but along the axes perpendicular to the side of the octagons needs to be calculated: $w_2 = w_1/\sqrt{[1 + (\sqrt{2} - 1)^2]} \approx 6.005″ \approx 6″$. This unit, as a whole number of inches, is, of course, simple to use. Use the axial lines of the square and octagons to mark the openings and turns, with the path width $w_2 = 6″$, as illustrated at right.

As a check on the above calculation of w_2, simply measure along one of the axes the distance between two successive octagons. The medieval master masons working geometrically could have set their compasses to this distance and then transferred this width to the openings. The above formula for w_2 and the explicit use of $\sqrt{2}$ would have been unknown to the medieval masons.

The more involved western entranceway and side will be accomplished in two more steps.

Step 8

Now, we will work on the north side of the western axis. The path width $w_2 = 6″$ is also the width of the opening of the labyrinth. Thus, a second line segment needs to be drawn 6″ to the north of, and parallel to, the western axis. Then the 6″ openings can be made as illustrated including the entranceway and final exit on the outer octagon.

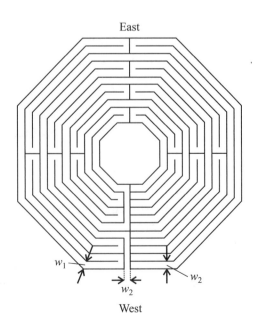

To concentrate our attention on the final product, the outer square has been removed from the explanatory picture. If you are using canvas, the light pencil markings can be left, as mentioned earlier, to show others the mathematical process underlying the labyrinth structure.

Step 9

Finally, we will work on the south side of the western axis. Another line segment needs to be drawn 6″ to the south of, and parallel to, the western axis. Then the 6″ openings can be made as illustrated, including the entranceway and exit from the small central octagon.

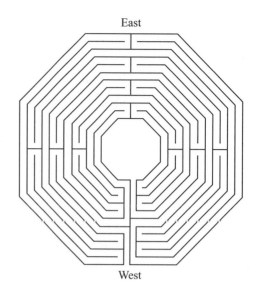

For a beige or off-white canvas labyrinth, once the design is laid out in pencil, the taping can commence with a better quality electrical tape. The completed canvas labyrinth can be folded or rolled for portability. Use again a double-sided carpet tape to hold the canvas taut in order for for people to walk on the labyrinth. People will need to remove their footwear to walk on the labyrinth in order to avoid stains to the canvas. Walking the labyrinth will provide a whole new geometric experience of the structure because the way in which the labyrinth is laid out is different from the way in which it is walked. For example, during the construction we did not put the turns in until near the end of the process, whereas a person walking the labyrinth encounters the turns throughout the walking process.

Insights of modern mathematics

Once you have constructed and walked the labyrinth, you may wish to stand back and learn more about its highly-ordered structure and more general mathematical features. The work by Leonhard Euler (1707–1783) on the theory of unicursal curves provides significant insights into the structure of mazes and labyrinths [11]. More recently, the study of topology has developed powerful insights for the study of mazes and labyrinths [12].

Euler's original motivation for examining unicursal curves arose from his celebrated solution and generalizations, first presented to the St. Petersburg Academy in 1736, on the question as to whether it was possible to take a walk from any starting point in the town of Königsberg and cross every bridge there once only and return to the starting point [13]. To understand some of Euler's insights some terminology is needed. First, a *node* is defined as a "point to or from which lines are drawn" [11, p. 245] and a *branch* "is a line connecting two consecutive nodes" [11, p. 245]. The *order* of a node is the number of branches that meet at it. An *odd* node is a node at which an odd number of branches meet or, in other words, a node with an odd order. Similarly, an *even* node has an even number of branches or an even order. A *route* is the "branches taken in consecutive order and so that no branch is traversed twice" [11, p. 245]. If the whole of a figure can be traversed in one route, then the figure can be described *unicursally*. Furthermore, a figure can be described unicursally if and only if the number of odd nodes is zero or two. Now, in terms of mazes, a dead end or blind alley in a maze would be an odd node, or more specifically, a node with order 1. Notably, a maze with only odd nodes at the entrance and center, and hence no blind alleys, can be traversed or 'walked' unicursally [11, p. 255]. A maze, in two or three dimensions, can be entered and exited by always following and keeping by the wall on the left-hand side, or similarly always keeping to the right-hand side [11, p. 258]. This route may not be the shortest route, but this procedure will insure that one is able to get out of the maze [11].

Further, for unicursal mazes leading to a center, "the path lies on a certain number of distinct levels" [14]. Level 0 is on the outside of the labyrinth, level 1 is the first layer (the first path width in from the outside), level 2 is the second layer, and so forth to the highest level number at the center.

Note, for example in the Amiens labyrinth [Figure 5], one starts, of course, at the outside, level 0, and then proceeds 5 path widths in to level 5, then level 6, then level 11, and so forth forming the level sequence 0, 5, 6, 11, …. Based on the order in which the distinct levels are encountered on the journey to the center, the level sequence, these mazes or labyrinths can be classified into distinct topological types and built up from elementary "submazes" [12, p. 65]. Topology is a field within mathematics that deals with the properties of geometric objects that remain the same under continuous deformation, and are independent of size and shape [15]. Thus, as we will note in Exercise Two, labyrinths with contrasting circular and octagonal shapes, may nevertheless have their turns in the same order, and hence be topologically equivalent. In other words, two labyrinths are topologically equivalent if one labyrinth can be transformed into the other labyrinth by a level-preserving deformation [12, p. 65]. Indeed, the strong constraints on the overall organization and size of unicursal mazes leading to a center, or labyrinths, result in relatively few topologically distinct patterns [12, pp. 65, 73]. In this manner, through the application of topology, the likely patterns of ancient Roman mosaic labyrinths have been reconstructed from the strong clues left in their surviving fragments [16].

Some class exercises

Try creating some ways to compare the structures of different labyrinths.

Exercise One. The labyrinth we constructed, based on the one at Amiens Cathedral [Figure 5], has 11 circuits, in the sense that the labyrinth walker makes, in sum total, 11 circuits around the centre before entering the central area. A person walking the Amiens labyrinth, of course, does not keep revolving in one continuous clockwise or counter clock-wise motion about the center. However, if one adds up all the segmented movements, between turns, on the journey to the center, the equivalent of 11 circuits are made. The simplest way to, in effect, see these 11 circuits in total is to observe the 11 'annular-like' spaces between the 12 concentric octagons illustrated in Figure 3 (which is similar to the illustration in Step 6, but without the diagonals). The labyrinth walker on the journey to the center (or from the center out) walks the entirety of all 11 of these 'annular-like' spaces or circuits.

Other numbers of circuits are possible and do, in fact, occur in other labyrinths. Do the labyrinths of Chartres and Reims [Figure 4] Cathedrals have 11 circuits? Try finding a picture or drawing of the Chartres labyrinth in an architectural history book or on the web.

Try drawing on letter-size paper a labyrinth with 4 or 5 circuits, so you become the designer of a simple labyrinth.

Exercise Two. Are the labyrinths of Amiens, Chartres and Reims Cathedrals topologically equivalent? That is, apart from differences in circular or octagonal form, do they have their turns in the same order? Note that the former labyrinth of Reims Cathedral [Figure 4] has intriguing octagonal 'pods' in four corners. Nevertheless, from the viewpoint of topology, can those roundabout paths be 'compressed back' into sides similar to those we constructed on our Amiens model?

Also, note that the three labyrinths of Amiens, Chartres and Reims Cathedrals have a similar structure based on two axes at right angles to each other. On the left-hand side of the Reims model, we note, proceeding from the left, that the outermost circuit is not blocked, but the next two circuits are blocked with turns, followed by one open, two blocked, two open, two blocked, and finally one open. Does this pattern hold on the other sections of this labyrinth or the corresponding sections of the other two labyrinths?

Learning from the Medieval Master Masons: A Geometric Journey through the Labyrinth

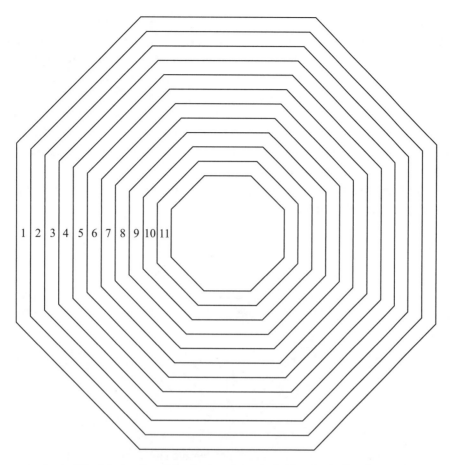

Figure 3. The equivalent of the 11 circuits of the Amiens labyrinth. Topologically, the numbers 1 to 11 also correspond to the levels of the labyrinth with level 0 the exterior of the labyrinth and level 12 the interior of the central small octagon.

Figure 4. Pattern of the former pavement labyrinth at Reims Cathedral.

Exercise Three. Compare the order and types of turns and path segments of the Amiens labyrinth we constructed [Figure 5]. Note that on entering the labyrinth and exiting the center, one first proceeds along a path segment with a length of 5 path widths before turning left! (On entering the labyrinth this length of 5 path widths is equivalent to passing along the levels 1 to 5 as illustrated in Figure 3. Similarly, on exiting the center the length of 5 path widths is equivalent to passing along the 5 levels of 11, 10, 9, 8 and 7.) For both of these cases of traveling in and out of the labyrinth, one next proceeds a quarter circuit, turns right and proceeds another quarter turn! Is there a continuing symmetry of movement here? Can you find a point in the labyrinth path where the half routes on either side of that point mirror or repeat each other? Is there a symmetry between the quadrants (in Figure 5, the lower left, lower right, upper right, and upper left quadrants)? The answers to these questions lead us to a deeper level in understanding the "secret of design" [17] or wonderful order underlying the Amiens labyrinth.

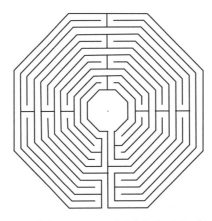

Figure 5. Pattern of the pavement labyrinth at Amiens Cathedral.

Exercise Four. Try drawing an 11-circuit labyrinth with its turns in a different order than those found at Amiens, Chartres and Reims Cathedrals. You may wish to use an enlarged photocopy of Figure 6 with some 'white out' to make the openings, and a pencil to create the turns and small line segments as needed. The 4 small protruding line segments on the exterior of the outermost octagon are guides for the labyrinth's perpendicular cardinal axes. This exercise can make you the designer of the unique pattern of an elaborate labyrinth! You may find that some patterns are more symmetric, interesting and ingenious.

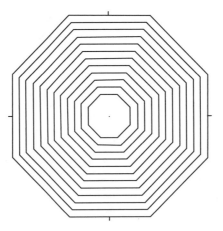

Figure 6. A basic concentric octagonal pattern for starting the design of an 11-circuit labyrinth.

Exercise Five. During our construction of the labyrinth, we noted that the length of half of the major diagonals of the outer octagon (e.g., AC in Step 5) measured from the square's center is $7\sqrt{[1 + (\sqrt{2} - 1)2]}' \approx 7.576745' \approx 7'\ 6\ 15/16''$. Using the Pythagorean Theorem, verify this expression and calculation. Can trigonometry also be employed to calculate the length of AC?

Exercise Six. In regard to the illustration for Step 8, we indicated in Step 7 that the path width w_2, measured not along AC (as in the case of w_1), but along the axes perpendicular to the side of the octagons, is $w_2 = w_1/\sqrt{[1 + (\sqrt{2} - 1)^2]} \approx 6.005'' \approx 6''$ where $w_1 = 6.5''$. Verify this expression and calculation. Hint: note that the expression $\sqrt{[1 + (\sqrt{2} - 1)^2]}$ also appears in Exercise Five.

The appearance of 6" and 6.5" relates closely to the modern carpenter's use of the ratio 12:13 (equivalent to 6:6.5) to construct an octagon rafter, a type of sloping wooden beam used in building some roofs [18]. What Pythagorean triplet includes 12 and 13? Explore how this Pythagorean triplet closely approximates the geometry involved in constructing an octagon.

Exercise Seven. Show that w_2's closeness to 6" arises from the fact that 407/288 is a very close rational approximation of the irrational $\sqrt{2}$.

Furthermore, during the construction of the labyrinth (Step 1), we checked the square's two diagonals in feet at $14\sqrt{2} \approx 19.7989\ldots \approx 19'\ 9\ 1/2''$, close to 20'. This calculation implies that 20/14 = 10/7 is a fairly close rational approximation of $\sqrt{2}$. Also, when we formed the outer octagon's vertices on the sides of the 14' by 14' square (Step 2), we noted that the octagon's vertices were close to 4' and 10' away from the two vertices of the square on the same side. Show that these observations arise from the fact that 14/10 = 7/5 is also a fairly close rational approximation of $\sqrt{2}$.

Theon of Smyra, c. 130 A.D., derived recursive formulae for generating rational approximations of $\sqrt{2}$ [19]. Find out about his procedure, and see if 7/5, 10/7, 14/10, 20/14, and 407/288 are generated by his method. In modern terms, Theon's method gives the same approximations, say x/y where x and y are natural numbers, generated by the continued fraction technique employed to solve the Pell equation in the form $x^2 + 1 = 2y^2$.

Conclusion

The design, construction and analysis of mazes and labyrinths involve wide-ranging areas in mathematics: Euclidean geometry and its constructions, arithmetic, trigonometry, the theory of unicursal curves, and topology. Most of this mathematics was, of course, unknown to the medieval master masons. However, their adeptness at practical geometry and mathematics was sufficient for the masterly design and construction of pavement labyrinths and the great church buildings that housed them. In determining a way for a winding path to fill the two-dimensional space encompassed by a labyrinth, these creative master masons were forerunners of modern topologists [20].

The cathedral and church pavement labyrinths had profound meanings associated with the human journey of overcoming fear and ignorance on the path of salvation [21]. The medieval master masons helped fulfill this purpose by the skilful application of geometry. Our own labyrinth work can likewise extend our knowledge of geometry, provide first-hand experience of the same mathematical issues faced by medieval craftspersons, and show how meaningful and applicable mathematics is in the history of art and cultures.

References

1. Lon R. Shelby, "The Geometrical Knowledge of Mediaeval Master Masons," *Speculum,* XLVII, 1972, 395–421.

2. Eric C. Fernie, "A Beginner's Guide to the Study of Architectural Proportions and Systems of Length," in *Medieval Architecture and its Intellectual Context: Studies in Honour of Peter Kidson,* Paul Crossley and Eric Fernie, eds., Hambledon, London and Ronceverte, West Virginia, 1990, 229–237. ———, *An Architectural History of Norwich Cathedral,* Clarendon Studies in the History of Art, Oxford University Press, Oxford, 1993. ———, *The Architecture of Norman England,* Oxford University Press, Oxford, 2000. Peter Kidson, "The Historical Circumstances and the Principles of the Design," in Thomas Cocke and Peter Kidson, *Salisbury Cathedral: Perspectives on the Architectural History,* HMSO, London, 1993, pp.35-98. ———, "Architectural Proportion, I. Before c. 1450", in *The Dictionary of Art,* Jane Turner, ed., Grove's Dictionaries, New York, 1996, 2: 343–352. Hugh McCague, "A Mathematical Look at a Medieval Cathedral," *Math Horizons,* X.4 (2003) 11–15, 31. Article available on the web at www.maa.org/news/Horizons-April03-McCague.pdf. Stephen Murray, *Notre Dame, Cathedral of Amiens: the power of change in Gothic,* Cambridge University Press, Cambridge, England and New York, 1996, 39–43, 170–173, Plate 45. We will be considering the geometry of the labyrinth at Amiens Cathedral which complements Murray's examination of the geometry of the ground plan of that building and the symmetry and structure of its labyrinth.

3. Nicola Coldstream, *Masons and Sculptors,* Medieval Craftsmen series, British Museum, London; University of Toronto Press, Toronto and Buffalo, 1991. This book is a very good overview of the methods, tools and materials of the medieval masons and sculptors. More detail on the medieval masons' instruments and methods appears in Lon R. Shelby, "Medieval Masons' Tools: The Level and the Plumb Rule," *Technology and Culture,* II (1961) 127–130. ———, "Medieval Masons' Tools II: Compass and Square," *Technology and Culture,* VI (1965) 236–248. ———, "Mediaeval Masons' Templates," *Journal of the Society of Architectural Historians,* XXX.2 (1971) 140–154.

4. The English Masonic Constitution of c. 1400 (Cooke MS., British Museum, Add.23198), thought to be largely based on an earlier mid-14th-century Constitution stresses the central role of "Geometry" in all things including stonemasonry. The text of the Cooke MS. appears in John H. Harvey, *The Mediæval Architect,* Wayland, London, 1972, 191–202 and in G. P., Douglas Knoop and Douglas Hamer, eds., *The Two Earliest Masonic MSS: The Regius MS. (B.M. Bibl. Reg. 17 A1), The Cooke MS. (B.M. Add. MS. 23198),* Manchester University Press, Manchester, 1938.

5. Craig M. Wright, *The Maze and the Warrior: Symbols in the Architecture, Theology, and Music,* Harvard University Press, Cambridge, MA and London, 2001. Penelope Reed Doob, *The idea of the labyrinth from classical antiquity through the Middle Ages,* Cornell University Press, Ithaca, 1990. Hermann Kern, *Through the labyrinth: designs and meanings over 5,000 years,* Translated from the German by Abigail H. Clay et al., Prestel, Munich, 2000. Kern's book presents the great range of labyrinths over the past five millennia, and is a good catalogue-like reference for information on specific labyrinths. An older, though still helpful, reference on labyrinths is: W. H. Matthew, *Mazes and Labyrinths,* reprint of Longmans, Green, New York, 1922, Dover Publications, New York, 1970.

6. Wright, *The Maze and the Warrior*, 41.

7. Wright, *The Maze and the Warrior,* 59–60. Stephen Murray, *Notre Dame, Cathedral of Amiens,* 78, 170.

8. Wright, *The Maze and the Warrior,* 50–56. Robert Branner, "The Labyrinth of Reims Cathedral," *Journal of the Society of Architectural Historians,* XXI.1 (1962) 18–25.

9. Robert Ferré, "Making A Masking Tape Chartres Labyrinth," Labyrinth Enterprises [updated c. 1995; accessed 19 July 2006]. Available from www.labyrinth-enterprises.com/tapec1.html.

10. John Carroll, *For Pros/By Pros: Measuring, Marking and Layout: A Builder's Guide,* Tauton Press, Newtown, CT, 1998, 16, 30–31. This book provides good practical advice on layout procedures from an experienced builder. The history of chalklines is also mentioned in "hand tool," *Encyclopædia Britannica,* 2004, Encyclopædia Britannica Online [accessed 24 November 2004]. Available from www.search.eb.com/eb/article?tocId=39233.

11. W. W. Rouse Ball and H. S. M. Coxeter, "Unicursal Problems," in *Mathematical Recreations and Essays*, 12th edn., University of Toronto Press, Toronto and Buffalo, 1974, 243–260.

12. Anthony Phillips, "The topology of Roman Mosaic Mazes," in *The Visual Mind: Art and Mathematics*, Michele Emmer, ed., MIT Press, Cambridge, MA and London, 1993, 65–73.

13. Ball and Coxeter, "Unicursal Problems," 243. Leonhard Euler, "The Seven Bridges of Königsberg" in James R. Newman, ed., *The World of Mathematics,* Simon and Schuster, New York, 1956, I: 573–580 is an English translation of Euler's later memoir on this subject. The generalizations of the Seven Bridges of Königsberg problem are sometimes referred to as Highway Inspector problems.

14. Phillips, "The topology of Roman Mosaic Mazes," 65. Phillips explains further these levels of a labyrinth in this article, and on his web site ---, "Through Mazes to Mathematics" [updated 6 April 2001; accessed 12 November 2005]. Available from www.math.sunysb.edu/~tony/mazes/index.html.

15. For example, a square and a circle are closed two-dimensional figures that are topologically equivalent. A square can be stretched into the form of a circle. Similarly, a donut (solid or filled-in torus) and a coffee mug are three-dimensional shapes each with only one hole, and thereby are topologically equivalent. A donut (solid torus) could be 'stretched' into the shape of a coffee mug without any cuts or punctures in the process.

16. Phillips, "The Topology of Roman Mosaic Mazes," 65–73. The reader is referred to this in-depth study of the topology and mathematical classification of ancient mazes, particularly Roman mosaic mazes.

17. Stephen Murray, *Notre Dame, Cathedral of Amiens,* 171. Murray, 171–173, provides a detailed analysis of the movements and symmetry of the Amiens labyrinth that were used for Exercise Three.

18. Thomas J. Emery and Frank D. Graham, *How to Use the Steel Square,* Classic Reprint Series, Algrove Publishing, Ottawa, 2001, 24, 26–27. The octagon scale on a rafter square, a steel square used to construct rafters, is one way that modern carpenters construct octagons. John Carroll, *For Pros/By Pros: Measuring, Marking and Layout: A Builder's Guide,* 50–51.

19. Peter Kidson, "A Metrological Investigation," *Journal of the Warburg and Courtauld Institutes,* 53 (1990) 71–97. Kidson discusses the method of Theon of Smyra in regard to such rational approximations applied in the history of measurement units, and mentions their applicability to the building crafts.

20. Phillips, "The Topology of Roman Mosaic Mazes," 73. Anthony Phillips actually describes ancient and medieval designers of mazes in even stronger terms as "ingenious, unsung topologists of the past" [p. 73].

21. Wright, *The Maze and the Warrior* [5].

Dem Bones Ain't Dead: Napier's Bones in the Classroom

Joanne Peeples
El Paso Community College

The year is about 1610; the place is Merchiston Castle near Edinburgh, Scotland; and the Baron of the castle, John Napier, is sitting at his desk working on his latest invention — logarithms. The computation of all of the tables of logarithms involved required many calculations, which had to be correct. Napier, ever the inventor, realized that his need to be able to calculate correctly, as well as the needs of many others at this time, made instruments that would mechanically compute of special interest. So … Napier picked up his bones (or rods as they were sometimes called) and started calculating.

Figure 1. Private collection of Joanne Peeples

It is not known when John Napier had his idea for "rod reckoning", a mechanical means to multiply, divide, square root and cube root numbers. He had started a manuscript about algebra and arithmetic, called *De arte logistica*, but put this work aside when he started working on logarithms (Napier's incomplete *De arte logistica* was not published until 1839). John Napier published *Mirifici logarithmorum canonis descripto* in 1614, with an English translation published in 1616. His next work was *Rabdology,* which was published in 1617. It is thought that the word *Rabdology* was a combination of Greek words meaning rod reckoning.

Napier's "bones" were not meant to replace paper and pencil, but to reduce the amount of time spent calculating and diminish errors in the calculations of products, quotients, square roots and cube roots. It is interesting to note that in *Rabdology*, Napier introduced the comma to separate the whole part of a number from the fractional part. Simon Stevin had used small circles in his 1585 work, but the comma was much easier to use. It took almost 100 more years before the comma/decimal point was in general use.

There is also controversy as to where Napier's ideas for his "rod reckoning" came from. Some people say that it was derived from a method the Arabs probably invented in about the thirteenth century and was brought to Europe during the Middle Ages. It was known as multiplication *per gelosia*, and was really multiplication on a grid [2, page 567]. This form of multiplication is illustrated below. To multiply 32 time 74 proceed as follows (this method assumes one knows single digit multiplication).

Make a two by two grid, putting one number on top, and the other on the right side (see Figure 2).

Insert the corresponding products in each square of the grid, with the tens digit above the diagonal and the units digit below as in Figure 3.

Add the diagonals, and read the answer starting at the top of the left side going down and across the bottom (see Figure 4).

The answer is 2,368.

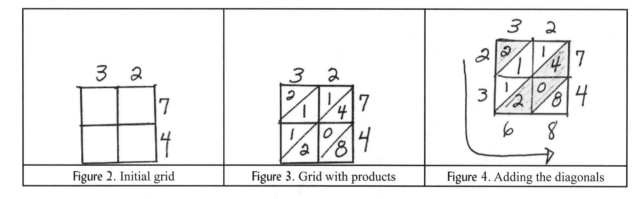

| Figure 2. Initial grid | Figure 3. Grid with products | Figure 4. Adding the diagonals |

According to Robin E. Rider [4, page xii] in his introduction to Richardson's translation of *Rabdology*, Napier used the multiplication table, and just inserted some diagonal lines. Looking at the *gelosia* grid, in hindsight the diagonals seem like a logical next step. Figure 5 is what such a table would look like (knowledge of single digit multiplication is not assumed)

As a person who used lots of numbers, and liked to solve problems, it would seem likely that John Napier "saw" the patterns, and in conceiving his rods used pieces of both the multiplication table (the columns of the table) and the multiplication *per gelosia* method (the diagonals on each entry in each column). John Napier included two other mechanical devices in his book, *Rabdology*. The first was the "promptuarium" which could be considered a precursor to the modern calculator (but will only multiply) and the other device looks like a checkerboard which performed calculations in binary (One wonders if Napier foresaw the computer).

Napier's rods were used well into the eighteenth century, with many variations. Often they were made of ivory, and from a distance looked like bones — hence, they are often called Napier's bones. There were bones whose cross-section was a square, there were bones that were round, with an entire multiplication table inscribed on each bone, and there were flat — paper bones (which were very popular since they could be easily included in a book).

NAPIER'S BONES

X	0	1	2	3	4	5	6	7	8	9
1	0/0	0/1	0/2	0/3	0/4	0/5	0/6	0/7	0/8	0/9
2	0/0	0/2	0/4	0/6	0/8	1/0	1/2	1/4	1/6	1/8
3	0/0	0/3	0/6	0/9	1/2	1/5	1/8	2/1	2/4	2/7
4	0/0	0/4	0/8	1/2	1/6	2/0	2/4	2/8	3/2	3/6
5	0/0	0/5	1/0	1/5	2/0	2/5	3/0	3/5	4/0	4/5
6	0/0	0/6	1/2	1/8	2/4	3/0	3/6	4/2	4/8	5/4
7	0/0	0/7	1/4	2/1	2/8	3/5	4/2	4/9	5/6	6/3
8	0/0	0/8	1/6	2/4	3/2	4/0	4/8	5/6	6/4	7/2
9	0/0	0/9	1/8	2/7	3/6	4/5	5/4	6/3	7/2	8/1

Figure 5. Multiplication table, with diagonal lines

Supposedly, in a speech Robert Hook made on May 7, 1673, to the Royal Society of London [4, page xvi], he concluded that pen and paper was preferable to any mechanical device for computation. Even back then, hand held calculators such as Napier's bones were controversial.

Following the order John Napier chose in *Rabdologia*, first I will describe how to make a set of bones, and then how to use them.

Making bones

The columns you will see on the various kinds of bones below are just the columns found on the multiplication table, shown earlier in this paper. For each entry of the table there is a diagonal line, from lower left to upper right, with the units digit below the diagonal and the tens digit above the diagonal.

The Napier's bones shown in the picture on the right were made by my student, Christopher Hoggard. He thought "bones" should be made of bone, and had the tools in his workshop to cut the rods from a large piece of real bone from an extinct Sea Elephant. The piece of bone he was working with was about 150 years old. Hoggard filled in the porous spots on the bones and had them inscribed using a laser. The result was a work of art.

Figure 6. From the private collection of Joanne Peeples

Another student, Susana Fernandez, used John Napier's instructions found in *Rabdologia* to construct a set of bones from wood (see Figure 1). She found some square dowels at a local hardware store, about 3/8 inch on each side, cut them into 3½ inch lengths, and using a fine tipped woodburning tool she inscribed the necessary numbers. Susana found that a harder wood (not pine) worked best.

Following Napier's instructions, the bones were inscribed in such a way as to be able to express all numbers less than 11,111 with one set of 10 bones. On each side of one bone, four different single digit multiplication tables were constructed, an example of the seventh bone is shown below (see Figure 7). I have "unfolded" the bone so you can see all four faces at one time, the 1, 4, 8, and 5 go on the ends of the rod, so you can easily see which bone to choose. Notice that the last two faces are upside down, also that 1 and 8 are nines-complements while 4 and 5 are also nines-complements.

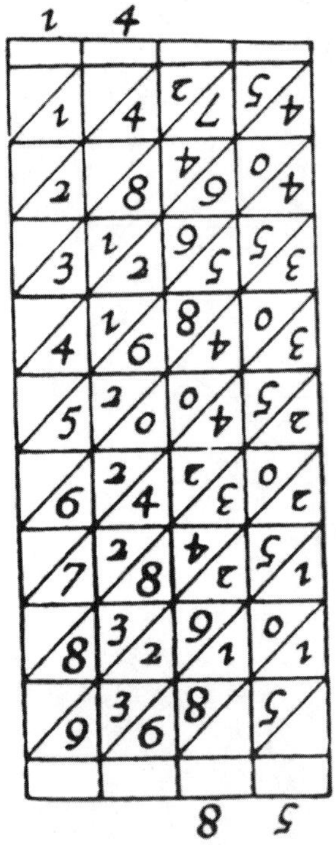

Each column in the diagram to the left (Figure 7) is a column of multiples of the number on the top — in other words just a column from the multiplication table.

The table below (Figure 8) shows which numbers to put on the bones.

Bones	Face I	Face II	Face III	Face IV
#1	0	1	9	8
#2	0	2	9	7
#3	0	3	9	6
#4	0	4	9	5
#5	1	2	8	7
#6	1	3	8	6
#7	1	4	8	5
#8	2	3	7	6
#9	2	4	7	5
#10	3	4	6	5

Figure 7. Faces of each bone

Figure 8. A sample of one bone.

Remember when making bones to flip the bone (end-to-end) before putting the numbers on Faces III and IV.

A much easier way to make a set of bones that can be used in the classroom is to make the bones out of tongue depressors. I have used these bones in 4th to 6th grade classes and with my liberal arts/preservice teacher classes. Figure 9 shows a set of my tongue depressor bones.

To make a set of these bones you will be marking on both sides of the tongue depressor, so your lines must be very accurate in order that everything will line up properly when using the bones to multiply or divide. I have made one line 1 1/16 inch down from the top and another 1 1/16 inch up from the bottom of the tongue depressor. The other eight lines are spaced 1/2 inch apart. The diagonal lines go from the lower left to the upper right in each of the nine rectangles. As you can see, one side of the tongue depressor seems to be "upside down" compared with the other side, and the "header" numbers (in this case 2 and 7) are nines-complements. Again, you can see that the numbers listed below the 7 and the 2 are just their multiples. You will need to make at least 10 bones, paired as in Figure 10.

Perhaps the simplest way to make a set of bones is to simply make a paper template like the one shown in Figure 5, make copies on a heavier paper, and cut the template copies into strips. The far-left column can

Dem Bones Ain't Dead: Napier's Bones in the Classroom

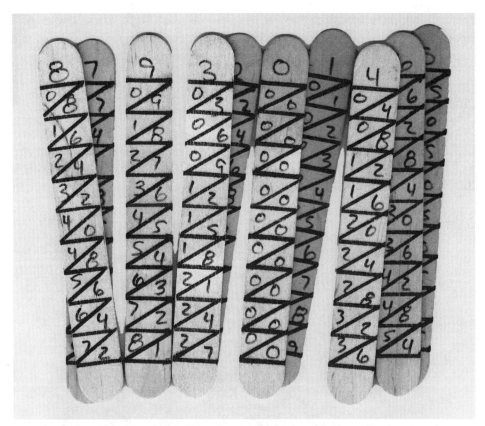

Figure 9. Tongue depressor bones from the private collection of Joanne Peeples

Side A	Side B
0	9
1	8
2	7
3	6
4	5
5	4
6	3
7	2
8	1
9	0

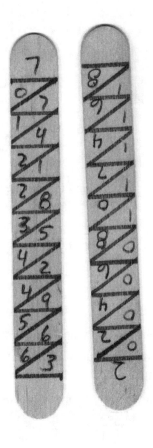

Figure 10. From the private collection of Joanne Peeples

be put on a flat surface to help count the number of rows you need to go down. These are not two-sided bones, but strips could be glued together (or glued onto tongue depressors) if you want to make them two-sided.

How to Multiply and Divide Using Bones:

Multiplying by a single digit number:

To multiply 47,526 by 7, first find the bones labeled 4, 7, 5, 2, and 6, and lay them side by side to make the number 47526. Then count down to row 7:

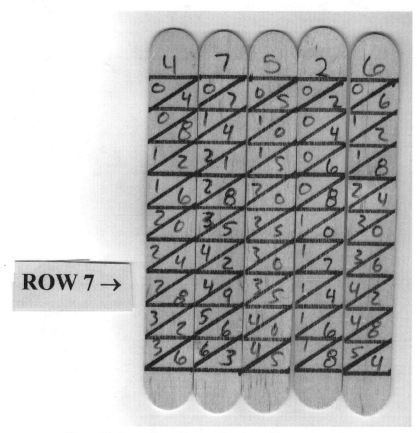

Figure 11. From the private collection of Joanne Peeples

By adding the numbers in the parallelograms (the first and last numbers are in triangles), you can "read off" the answer.

The ones-digit is (in the triangle to the right) 2.

The tens-digit is (in the first parallelogram) is 4 + 4 = 8.

The one hundreds-digit (in the second parallelogram) is 5 + 1 = 6.

The one thousands-digit (in the third parallelogram) is 9 + 3 = 12, so write down the 2 and carry the 1.

The ten thousand digit (in the fourth parallelogram) is 8 + 4 + the 1 that was carried = 13, so write down the 3 and carry the 1.

The last digit (in the triangle on the left) is 2, add the 1 that is carried and you have 3.

Your number is: 332,682.

Multiplying by a number with more than one digit:

To multiply 615 by 365 you will need some paper. First set up 615 on your bones. Then on your paper write 365 with a line under it. Then multiply 615 by 5 using the same method as above, then multiply by 6, and finally by 3, writing your results as follows in Figure 12:

<u>365</u>

			3	0	7	5
		3	6	9	0	
	1	8	4	5		
	-	-	-	-	-	-
Adding we get	2	2	4	4	7	5

Figure 12. Multiplication

Your answer is 615 × 365 = 224,475.

Multiplying — a second method:

This method uses nines-complements. To multiply 615 by 365 using nines-complements, set up 615 on your bones. Then "flip" (turn them over, end over end) your bones, and you should have 384. You will then want to multiply 384 (the nines-complement of 615) by the number 365. First, multiply 384 by 5, then by 6 and finally by 3, writing your results as follows in Figure 13.

<u>365</u>

			1	9	2	0
		2	3	0	4	
	1	1	5	2		
Write 365 in here				3	6	5
	--	--	--	--	--	--
Adding we get	1	4	0	5	2	5

Figure 13. Multiplying using nines-complements

Now subtract 140,525 from 365,000 (adding enough zeros to 365 to make as many digits as in your sum) as in Figure 14.

		3	6	5	0	0	0
	-	1	4	0	5	2	5
		-	-	-	-	-	-
Subtracting we get		2	2	4	4	7	5

Figure 14. Second step in multiplication using nines-complements

Your answer is 615 × 365 = 224,475, the same as was found by the first method of multiplication.

Division:

To divide 589,475 by 365 complete the following steps:

Step A. Write 589,475 on your paper and set up 365 on your bones. Among all the multiples of 365 that are less than, but closest to 589 is 1 (1 × 365 = 365, 2 × 365 = 730 which is too big). So 365 should be written under the 589 and subtracted from 589. The difference, 224, is to be written above the 589. And finally the number 1 is written as part of the quotient (see Figure 15).

Step B. On the sixth row of your bones you will find 2190, which is 6 times 365, and is closest but less than 2244 (the last 4 in 2244 was taken from 589475, as we do in normal long division — and I have italicized it in the diagram below). So write 2190 at the bottom, and the difference between 2244 and 2190 at the top — as indicated in Figure 16.

Step C. Repeat step B, this time looking for the largest multiple of 365 that is less than 547. See Figure 17.

Step D. Repeat step B again, this time looking for the largest multiple of 365 that is less than (or equal to) 1825. See Figure 18.

Step 2	2	2	4					
	5	8	9	4	7	5	(	1 Step 3
Step 1	3	6	5					

Figure 15. First step for division

Step 5			5	4					
		2	2	4	*4*				
		5	8	9	4	7	5	(	1 6 Step 6
		3	6	5					
Step 4		2	1	9	0				

Figure 16. Second step for division

Step 8			1	8	2					
			5	4	7					
		2	2	4						
		5	8	9	4	7	5	(	1	6 1 Step 9
		3	6	5						
		2	1	9	0					
Step 7			3	6	5					

Figure 17. Third step for division

Step 11					0					
			1	8	2	*5*				
			5	4						
		2	2	4						
		5	8	9	4	7	5	(	1	6 1 5 Step 12
		3	6	5						
		2	1	9	0					
			3	6	5					
Step 10			1	8	2	5				

Figure 18. Division is complete

Your quotient is 1615.

After some practice, division becomes quite easy as long as you keep your columns nice and straight. Also, your subtraction skills will improve.

Where can these bones be used?

One place where I have found them effective is in 4th–6th grade classrooms. Both the teachers and the students have been enthusiastic about learning to use the bones. Two quotes from students:

"I really enjoyed learning about how to use the Napier's Bones. I study them all night. P.S. I want to go to college to train as a teacher."

". . . I learned a lot of things, like Napiers Bones. Napiers Bones was cool because I was learning math and history at the same time. I could multiply on sticks. It sounds weird but fun."

Napier's bones were used in the elementary classroom as late as the 1960s [1] in England, to help teach multiplication.

Another place to use the bones is in the lowest level of developmental math classes, often taught at community colleges. When adults have trouble multiplying and dividing, a good learning tool is to expose them to a different way of doing things. They can even check their answers on a calculator — and they often find it amazing that they get the same answer.

The place I like using them the best would be in a liberal arts or preservice teacher class. The students make their own set and learn to multiply and divide using "their bones".

I've had students make bones for other number bases, such as base 5 bones, and do calculations with them. There are references [3] that indicate Jonas Morre (1627–1679), of England, and Samuel Reyher

Figure 19. Students in a math-education class at El Paso Community College (Private collection of Joanne Peeples)

Figure 20. Students taking an exam at El Paso Community College (Private collection of Joanne Peeples)

(1635–1714), of Germany, may have used base 60 bones when making astronomical calculations, since degrees, minutes and seconds can be thought of as a base 60 system. Depending on the class, the question could be posed, "Why does the "nines-complement" multiplication work?" Also, the students could be taught to take square roots and cube roots (for this I suggest looking at Napier's *Rabdologia*, I have found the instructions there less confusing than the instructions I have found searching the web).

Conclusion

Students of all ages enjoy making and practicing multiplication on their bones. For the elementary student, it adds a kinesthetic element to mathematics, which is frequently absent. For the developmental student in the community college setting (some of whom still have trouble multiplying), the bones provide a different way of "looking at" multiplication.

In the Liberal Arts math classes at my college, we look at different number systems. After teaching the Babylonian number system, the Mayan number system, and we start looking at our number system more closely, I find that using the base 10 bones causes the students to think more carefully about how to compute with our numbers. Also, before using base 5 bones, the students had all sorts of problems multiplying in base 5 — using the bones, almost all the students are able to get the problems correct. I think the hands-on experience of making and calculating with concrete objects adds another dimension to the students' understanding.

In the Preservice teacher class I use the bones in multiple ways. Since these students are going to be grade 8 or lower teachers, they now have another "tool" in their bag of tools to use in the classroom – another way to show their students multiplication. In this class we also learn to divide using the bones. In learning division, the students see just how similar it is to the long division they are used to doing, but the numbers are placed in different spots on the page. It provides a non-standard way to do a standard problem, which I find good for their thinking skills. Also, when they multiply using the nines-complement method they have to change their method of thinking. If the mathematics requirement for this Preservice teacher course was higher at my college, I would have them look at "why" the nines-complement method works.

I am still exploring more ways to use the bones in my classes. I have always used bones in my liberal arts and education classes, but because of the positive reaction from students a set of bones is now included in our basic math textbook.

References

1. Diploudis, Alexandros, www.cee.hw.ac.uk/~greg/calculators/napier/about.html.
2. Ifrah, Georges, *The Universal History of Numbers*. Published by John Wiley and Sons, Inc., 2000.
3. Jones, Phillip S., Tangible Arithmetic I: Napier's and Genaille's Rods, *Mathematics Teacher* 47 (Nov. 1954), 482 – 487.
4. Napier, John, *Rabdology*, translated by William Frank Richardson. Volume 15 in the Charles Babbage Institute Reprint Series for the History of Computing. Published by the Massachusetts Institute of Technology and Tomash Publishers, 1990.

The Towers of Hanoi

Amy Shell-Gellasch
Pacific Lutheran University

Introduction

The classic Towers of Hanoi puzzle (Figure 1) is known to many of us, either from our school days or as educators. The puzzle works as follows: given some number of discs (washers) of varying diameters stacked in a pyramid on one of three posts, move the stack one disc at a time to one of the other posts in as few moves as possible. The catch is that no disc can rest on top of a disc of smaller diameter. This is a wonderful problem for school age students and many people first encounter it in elementary or middle school. Though the puzzle can be presented as a thought exercise (without the physical model), having the physical puzzle enhances discovery, and as such is an essential aide in any course that discusses dynamical systems. Outside of this venue the puzzle can be used in courses from elementary school through college to teach reasoning, problem solving and the development of mathematical formulas. It can be used to discuss recursion and induction as well as graph theory with students at the secondary and college level.

The Towers of Hanoi puzzle

Given three pegs and any number of discs of increasing radius stacked in a pyramid shape on the first peg, move the stack to the third peg. Only one disc may be moved at a time and no disc of larger radius may be placed on top of a disc of smaller radius. What is the smallest number of moves to move the stack?

Figure 1. Towers of Hanoi, Constructed by Bill Acheson.

Generalization: given *n* discs, find the minimum number of moves to move the stack given the above restrictions.

Solution

In the trivial case of one disc, one move is needed to move the disc.

In the simplest (non-trivial) case of two discs, move the top (smaller) disc temporarily to peg two in the middle, move the second (larger) disc to the final peg (peg number 3), then the first disc is placed on top of the second disc, thus completing the move. So to move a 2 disc pyramid, three moves were required.

In the next case, consider a 3-disc tower on peg number one (on the left). For simplicity, we will label the discs 1, 2, 3 by increasing size (thus disc 1 is the top disc). To move the stack to peg number 3, do as follows:

Move disc 1 to peg 3.
Move disc 2 to peg 2.
Move disc 1 on top of disc 2.
Move disc 3 to peg 3.
Move disc 1 to peg 1.
Move disc 2 to peg 3.
Move disc 1 to peg 3.

The optimal number of moves to move three discs is seven. As you see, to complete this move, we first moved the top two discs (a smaller stack) to the middle peg as in the previous case, then moved the bottom disc to the final peg, and then moved the smaller stack to the final peg. Thus this is a recursive process.

So, to move a 2-disc stack requires three moves and to move a 3-disc stack requires $2 \times 3 + 1$ or seven moves. To move a 4-disc stack requires fifteen moves. The table below can be filled in by students to help them see the recursive relationship.

n	# moves	formula
1	1	1
2	3	3
3	7	$2 \times 3 + 1$
4	15	$2 \times 7 + 1$
5		

In general, to move *n* discs, move the top $n - 1$ discs to the middle peg, move the bottom disc to the final peg, then move the top $n - 1$ discs now on the middle peg to the final peg. If we let $a(n)$ represent the number of moves to move *n* discs, the recursive formula is

$$a(n) = 2a(n-1) + 1.$$

Generalization to a non-recursive or analytic solution can be introduced at this point. Induction can be used to verify the student's hypothesis. The analytic solution to our recursive formula is $a(n) = 2^n - 1$.

History

In the larger scope of the history of mathematics and its devices, the Tower of Hanoi has a relatively short history, though the non-mathematical basis for the puzzle is much older. An "ancient relic", as David Poole describes it, exists in a Buddhist temple near Hanoi, Vietnam. [3]

It is doubtful that this obscure artifact elicited the interest in the puzzle that was prevalent in the nineteenth century and led to the first construction of the physical puzzle in 1883. The source of this interest is most likely the mythical story of the Tower of Brahma. In this tale, a group of Brahmin monks have three diamond needles and 64 golden discs. The story foretells that the world will end when the monks have moved the pyramid using the above rules at the rate of one move per second. (Ask your students if they will see that day.[1]) Thus, this tale is also referred to as the "end of the world" problem.

Recreational mathematics surged in popularity in the 1800s, and the Tower of Brahma story was popularized in the writings of Frenchman H. de Parville. [3] The first (known) physical puzzle, now commonly known as the Tower of Hanoi, was constructed in 1883 by French number theorist François-Édouard-Anatole Lucas. [1] The first published solution appeared the following year by M. Lucas in *La Nature* under the pseudonym M. Claus. [4] Since that time, the Tower of Hanoi has been a favorite of recreational mathematicians the world over. Ian Stewart cleverly explores the connection between the Towers of Hanoi puzzle, graph theory and Sierpinski's Triangle in [4].

Construction

Construction of this model is fairly simple. The simplest solution would be to go to an educational store, the ones that have all kinds of scientific and mental games and devices, and purchase one. If you are more interested in making one from scratch, all the materials should be available at your local home improvement store. First decide the dimensions of your model. (The nicely finished one shown above measures $30 \times 10 \times 10$ cm or $12 \times 4 \times 4$ in, while the more utilitarian one below for classroom use is larger.) You will need a piece of one- or two-inch thick board for the base, three dowels for the posts, and discs of increasing radius. Six or seven discs will suffice for classroom use. If wooden discs are not available, buy half-inch plywood and use a jigsaw to cut your discs. Drill holes in the center of each disc large enough for the dowels to slide easily through. Drill three holes on the base (spaced so the largest two discs do not interfere with one another) large enough for the dowels to fit into snugly. Use a little wood or Elmer's glue to fix the dowels in place. Finally sand all edges to avoid splinters.

The result will look something like Figure 2.

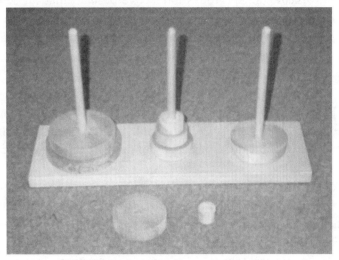

Figure 2. Simple classroom model

Since this rough and ready version may only require a drill and glue (if pre-cut discs can be bought), groups of students may be interested in building their own. Do however make sure that an adult is aware of the project to avoid any issues involving power tools. For a nicer version, round off all edges and stain and

---
1 This is on the order of 1.8×10^{19} years.

varnish to your liking. The model shown at the beginning of this article was constructed by Bill Acheson[2] and proudly sits on my desk. For a really quick and inexpensive version that students can make in class, use cardboard for the discs, a shoe box lid for the base, and pencils for the pegs. This version has the benefit of being able to be easily dismantled.

In the classroom

This model can be used at almost any level. At the simplest level, school age students can use the model to experiment with finding patterns. You may even want to devise other sets of rules for moving discs besides the ones associated with the traditional Tower of Hanoi problem. For example, you can suspend the rule of no larger disc on top of a smaller disc and ask how many moves are needed to move the pile. (The answer is $2n$, invert the tower onto peg 2 and then invert it again onto peg 3.) This is a good initial problem that then leads into the full problem with the size restriction. Providing a blank table of *number of discs* versus *number of moves* as shown earlier will help facilitate pattern discovery and writing the rule algebraically. You may also want to have a discussion of the growth of powers; how big is 2^n? The Legend of Sessa, in which a mathematician is paid by placing one grain of wheat on the first square of a chess board, two on the second, four on the third, etc., is a nice story to use to get them thinking about size. [2] The full Tower of Hanoi problem can be given to students in middle or high school classes. Using this exercise is a wonderful way to introduce recursion, and ultimately build up to the notation of recursive formulas. At this level, students can then experiment with their own rules and see if their version leads to a recursive formula also.

At the college level, the Tower of Hanoi can be used to introduce induction in almost any course. It can be used to study recursion and all its formalities in a discrete or continuous dynamical systems course. In the freshman level discrete dynamical systems course in which I used this model, it was presented on the first day of class to introduce recursion and recursive formulas to set the groundwork for much of the semester. A set of six to eight towers of Hanoi were passed out to the class.[3] After the rules of the puzzle were explained, students worked in small groups to discover a mathematical description for the number of moves needed to move a pile of n discs. As I walked around, I encouraged them and asked directed questions to keep them on track. Most groups will quickly try the problem with first two or three discs and then work up from there. If a group does not notice that line of attack, suggest it. After approximately ten minutes, most groups will have a formulation of the answer. Group discussion can bring out the rule in general terms. Then a Socratic method can be employed to introduce recursive notation (usually the $a(n)$ and $a(n-1)$ notation). From that point other situations can be explored (with or without the model) to generate other recursive formulas. One of the goals of studying recursion is the notion and accompanying notation of long term behavior and eventually the study of numerical and analytic solutions of differential equations.

Much more can be explored using the Towers. In a computer science course, it can be used to discuss the differences between iterative and recursive algorithms. For the applications of the Tower to graph theory (Hamiltonian paths), linear algebra, number theory, group theory and Pascal's triangle (in the form of odd binomial coefficients) see Poole [3]. The relationship between the puzzle, graph theory and Sierpinski's Triangle is explored in [4]. For a discussion on the tower and its relationship to formal language theory (as morphisms, their fixed points, and as square free strings) see Allouche, et. al. [1].[4]

---

[2] See the article on string models in this volume by Acheson and Shell-Gellasch for more of Bill's handiwork.
[3] This was while I was teaching at the United States Military Academy at West Point, New York. The Department of Mathematical Sciences has about two dozen of the simply constructed models (as described above) for classroom use.
[4] References 1 and 3 are available on JSTOR.

Conclusion

The Tower of Hanoi is an intriguing problem for students of all levels and abilities. It can be used in the classroom from elementary school to college as well as a range of topics from algorithms to graph theory. It is a wonderful motivator, providing some history while engaging problem solving skills and provides a tactile connection with mathematics.

References

1. Allouche, Jean-Paul; Astroorian, Dan; Randall, Jim; Shallit, Jefferey; Morphisms, Squarefree Strings, and the Tower of Hanoi Puzzle, *American Mathematical Monthly*, **101** (7), Aug-Sept., 1994, pp. 651–658.
2. Ifrah, George, *The Universal History of Numbers*, John Wiley & Sons, 2000, pp. 323–324.
3. Poole, David G., The Towers and Triangles of Professor Claus (or, Pascal Knows Hanoi), *Mathematics Magazine*, **67** (5), Dec., 1994, pp. 323–344.
4. Stewart, Ian, *Another Fine Math You've Got me Into…*, Dover Publishing, 2003, pp. 5–14.

Rectangular Protractors and the Mathematics Classroom

Amy Ackerberg-Hastings
University of Maryland University College

Introduction

It may not occur to contemporary students that architects, surveyors, and drafters once did their work by hand, without the aid of computers. The nineteenth-century drawing instruments used to prepare engineering drawings may seem unusual in their form and materials to a twenty-first century audience. Meanwhile, although they likely used protractors to measure and draw angles in middle school and learned a general proof for bisecting the angle in high school, even mathematics majors may never realize that someone had to divide the circle precisely in order to manufacture instruments that measure angles. The rectangular protractor was a fixture of the engineering toolbox in the nineteenth century (Figure 1). Replicating this device is a convenient way of introducing abstract processes for drawing angles of a certain number of degrees to students in preservice teaching courses and perhaps also at the secondary level. This chapter provides the historical context for rectangular protractors and instructions for making them.

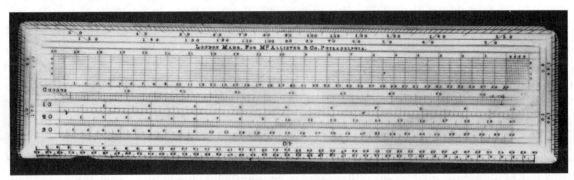

Figure 1. Ivory rectangular protractor sold by William Y. McAllister, c. 1836–1853. Negative number 72-9870. Catalogue number 310743. Courtesy of the Smithsonian's National Museum of American History.

Description and history

A rectangular protractor is simply a device for laying off and measuring angles in which the angle marks are displayed around the three sides of a rectangle instead of along a semicircular arc. English mathematical practitioners invented the protractor at the end of the sixteenth century to aid navigators and surveyors who needed to plot sea courses or to make technical drawings of land and buildings. [1] Although the first protractors were circular in shape, instrument makers throughout Europe began to provide other formats over the next several decades. One displayed the angle marks along the semicircular (180-degree) arc used to teach angle measurement in American elementary and middle schools today. Another typical seventeenth-century style looked like what modern readers might think of as a ruler. (See Figure 2.) Early brass versions of these rectangular protractors were often nearly as wide as they were long, following a design that appeared in the second edition of William Leybourne's *The Compleat Surveyor* (1657). [18] In the eighteenth century, English makers also made narrower rectangular protractors from silver or from ivory, establishing the style depicted in Figure 1. The degree marks on all of these instruments would have been engraved by hand against a master pattern. [8, pp. 188–204]

Figure 2. English surveyor's rectangular protractor, made of brass, c. 1660, based on the design in the second edition of *The Compleat Surveyor* by William Leybourne (1657). Inventory no. 40226. By permission of the Museum for History of Science, Oxford.

The rectangular protractor reached its height of popularity in the middle of the nineteenth century. Makers of drawing instruments then routinely sold rectangular protractors made of ivory or boxwood that were usually six inches in length and almost two inches wide. They were able to produce protractors inexpensively and in relatively large quantities because the degree divisions were now marked by machine. The rectangular protractor was marketed to drafters and mechanics as a versatile, convenient tool that could fit in a pocket. In addition to selling these instruments separately, makers often included rectangular protractors in boxed sets of drawing instruments. A young man in the process of mastering the art of mechanical drawing might purchase or be given such a set for use throughout his working life. He would use the instruments to develop skills at drafting through self-study or through apprenticeship in a machine shop. Alternatively, and especially if he was raised in an upper socioeconomic class, the owner of a set of drawing tools might study architecture and surveying more formally in a military academy.

During the nineteenth century, most makers of rectangular protractors lived in England. Although American practitioners and instructors often added rectangular protractors to their toolkits as well, they had to purchase English-made instruments imported by American dealers. For instance, in 1867, the Philadelphia optical firm operated by William Y. McAllister retailed "London made" ivory rectangular protractors of six inches for $1.50 and of twelve inches for $12.00. The six-inch protractors were marked with an additional thirteen scales, while the twelve-inch versions carried twenty-two extra scales. [17, pp. 24–25]

Indeed, throughout the nineteenth century, makers normally placed numerous scales on the interior of the rectangular protractor in order to make the instrument seem yet more useful to prospective customers. These scales provided little practical help in preparing engineering drawings since they were difficult to read accurately when they were not along the edge of an instrument. Nonetheless, it was not uncommon for most or all of the following scales to appear on rectangular protractors:

1) Rulers in inches and/or centimeters.

2) Scales for logarithmic calculation (cosines, sines, tangents, semi-tangents, lengths of chords of arcs from 0° to 90° in a circle of unit radius).

3) Architect's scales (dividing the inch into 10–60 equal parts, 1/8- to 1-inch to the foot). These scales enabled architectural drawings to be read directly in feet.

4) Scales for surveying plans (chain scales, the diagonal scale). Surveyors carried heavy metal chains with 100 links to locate property bounds. A chain scale was used to prepare scaled drawings of these measurements. A diagonal scale was used to find subdivisions within the various units of measure. [13]

5) Scales for navigational calculation (rhumbs, meridian line, latitude, longitude, hours, tides, masthead).

6) Military scales (proportions).

In the twentieth century, Americans were finally able to buy rectangular protractors produced domestically by the New York instrument manufacturer, Keuffel & Esser (K & E). For example, in its 1909 catalog, K & E listed a six-inch ivory rectangular protractor with eleven extra scales for $1.60 and rectangular protractors that were twelve inches long and contained twenty-five extra scales for $11.50. [4, p. 176] By World War II, Americans could also purchase transparent plastic rectangular protractors that were manufactured by American companies such as K & E or Felsenthal & Sons of Chicago. These instruments were used for positioning artillery or for aeronautics.

The rectangular protractors described above are historical artifacts today. Examples are regularly sold by rare instrument dealers or on Internet auction sites. Expect to pay $20.00-75.00, depending on the condition of the protractor. While these protractors were rendered obsolete for mechanical and surveying applications by computer-aided drafting, vinyl record enthusiasts still use a version of the rectangular protractor to align their turntables. Additionally, some sailors and pilots continue to employ square protractors made from plastic or paper as they navigate the sea or air. Examples of these may be found at Web sites catering to these activities, such as Flightstore: The Flight and Model Store. [10] Finally, the Fisher Scientific Company still sold a relatively inexpensive, clear, plastic rectangular protractor in sets of twelve as of early 2007. [9, item number S40651]

Making a rectangular protractor

An instructor who wants to emphasize angle division can plan a workshop on constructing rectangular protractors. Few materials are required: tagboard, compasses, straightedge, drawing pencils, and scissors for each student. Depending on the academic maturity of the students and the instructor's objectives, rulers and professionally manufactured, semicircular protractors may also be needed.

The basic procedure for making a rectangular protractor is the same, regardless of the amount of time allotted or the sophistication of the circle-dividing techniques employed. First, ask students to draw a semicircular arc of three-inch radius with a compass on tagboard. Second, they should divide the arc to the instructor's desired level of accuracy. Third, direct them to superimpose a rectangle with a length of six inches and width of two inches upon the arc. (See Figure 3.) If necessary, students can extend the angles they have drawn to the edges of the rectangle. Finally, they should cut out their rectangular protractor and experiment with using it to measure and lay off angles.

With elementary or middle school students, an instructor may wish simply to enrich the daily routine of class with an illustrated presentation on changes in engineering and surveying practice and a few minutes

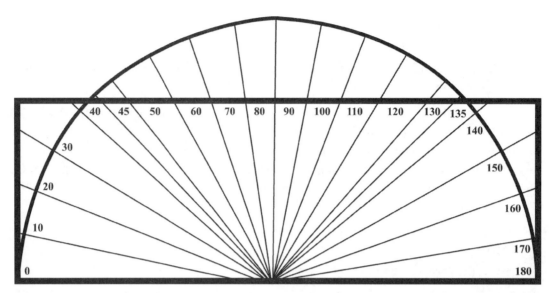

Figure 3. How a divided semicircular arc with the form of a rectangular protractor superimposed might look. The students would mark angle values along the inner edge of the rectangle and then cut out the rectangle. Image created by Jennifer Nugent.

spent drawing angles on a rectangle with the aid of a commercially-made semicircular protractor. The pupils then could compare measurements made with their rectangular protractors against the readings on their semicircular instruments. A high school teacher could use this project to provide practice in repeatedly bisecting angles. The resulting rectangular protractor would be graduated into the following increments of degrees: 45/64, 1 13/32, 2 13/16, 5 5/8, 11 1/4, 22 1/2, 45, 90. As discussed below, high school geometry or trigonometry classes could also make rectangular protractors in part to motivate an exploration of angle trisection.

The construction of a rectangular protractor may be taken to greater depths by those professors and teachers who are dismayed that the topic of angle division is omitted from many mathematics textbooks in current use. (For middle or high school texts that employ discovery-type activities yet lack information on how specific angles might be drawn without a protractor, see, for example, [7] [12] [21] [22]. In contrast, at least one "new math" textbook taught fifth-graders to create a protractor graduated into eighths; [20, pp. 702–721].) Preservice teachers, mathematics majors, or high school students might be challenged to try to make a rectangular protractor on their own at first. As they come to the realization that laying out angles is a non-trivial exercise, the instructor might suggest that thirty-, forty-five, and sixty-degree angles can be drawn with a compass or by constructing right triangles. In addition, students could use trigonometry to construct triangles with one- and ten-degree angles, translating these angles to the arc to fill in intervals of ten degrees and/or of one degree.

Alternative, detailed discussions of techniques for graduating an angle appear in J. L. Heilbron's *Geometry Civilized*, [11, pp. 90–97], and among the high school geometry frequently asked questions on "Ask Dr. Math." [16] The reader may also consult older textbooks in practical geometry or the excerpt from Thomas Hill's *A Second Book of Geometry* (1863) reprinted in "Protractors in the Classroom: An Historical Perspective." [1]

Indeed, the process of making a rectangular protractor demonstrates that dividing the circle is a significant problem in the history of mathematics. It dates back to the straightedge and compass constructions of Euclid's geometry and the impossibility of trisecting the angle with ancient Greek tools alone. [3; 16] Despite its difficulty, however, mathematicians and scientists viewed angular measurement as the key to the theory and practice of astronomy and navigation until the twentieth century. Therefore, scholars and practitioners have historically searched for ways to improve the precision of their measurements. They devised methods for dividing the circle analogous to the techniques described in the previous paragraph and invented mechanical devices—screws, verniers, and small microscopes or magnifiers—for reading instruments to higher levels of accuracy. They applied these precision modifications to simple instruments (including specialty protractors) as well as to complex devices (such as astrolabes, sextants, theodolites, and astronomical circles). For instance, a protractor without a vernier (a sliding scale for indicating parts of the divisions on a graduated instrument) might allow measurements within one-half degree (30 minutes). With a vernier, the user might be able to take a reading to five minutes of accuracy. The history of improvements in precision shifted between application and theory, as craftsmen and scientists pushed each other to new observational achievements. Their mutual give and take ran its course only with the advent of optical tools and electronic equipment for astronomy and of digital computers for navigation. [2; 5]

Conclusion

As mathematical instruments, rectangular protractors were inherently limited. Although makers marked them with numerous scales so that they appeared versatile, these scales had little practical value since most of them were in the interior of the protractor rather than along its edge. As with other shapes of protractors, therefore, these instruments were chiefly confined to transferring angular measurements to paper and to laying off angles for architectural and engineering drawings. [**8**, pp. 8-26] Another drawback to rectangular protractors made of ivory or boxwood was that these materials would shrink with age, rendering incorrect the angular divisions. When a precise angle measurement was needed, then, practitioners generally chose semicircular brass protractors, which gave more precise readings and which were stable over the long term. Still, since rectangular protractors were inexpensive and pocket-sized, they were adopted somewhat successfully outside the workshop as educational tools for teaching about angles and the various scales. Rectangular protractors thus fit within an international pedagogical trend in the nineteenth century toward an "informal geometry" based on experiment and intuition. [6; 23] Even there, though, rectangular protractors form only a small part of the history of geometry education. (For a brief but comprehensive account of the history of American mathematics education, see the online exhibit, "Slates, Slide Rules, and Software." [15])

Yet, despite their relatively minor historical importance, rectangular protractors are worth incorporating into college or school mathematics instruction for a variety of reasons. These instruments may stimulate students' interest in the geometry of angle division. Making their own rectangular protractors can help those mastering mathematics and mathematics education to evaluate the challenges involved in dividing the circle. This activity—and thinking about the limitations on the accuracy of rectangular protractors—additionally guides potential teachers as well as their students toward the 2000 NCTM standards of "establishing the validity of geometrical conjectures using deduction, proving theorems, critiquing arguments made by others; using trigonometric relationships to determine lengths and angular measures" and "analyzing precision, accuracy, and approximate error in measurement situations." [19, pp. 309, 321] Additionally, the form of rectangular protractors is unexpected by most contemporary American students, so a presentation on the instruments can catch their attention. In turn, this could lead to research into the different "standard" shapes for protractors found around the world. For instance, the counterparts to American high schoolers in the Netherlands use protractors in the shape of right triangles. Finally, students can benefit from exposure to some of the practical concerns represented by the various scales found on rectangular protractors, such as the effort nineteenth-century drafters put into reducing and enlarging mechanical drawings. Angle division is a vital subject in geometry and trigonometry. Building a lesson around the construction of a rectangular protractor is one way the mathematics teacher can emphasize this topic without a burdensome amount of advance preparation.

Acknowledgements

The author is indebted to Peggy Kidwell, the Smithsonian's National Museum of American History, and Jennifer Nugent for assistance with the preparation of this paper.

References

1. Amy Ackerberg-Hastings, Protractors in the Classroom: An Historical Perspective, in *From Calculus to Computers: Using the Last 200 Years of Mathematical History in the Classroom*, ed. Richard Jardine and Amy Shell-Gellasch, MAA Notes No. 68, Washington, DC, 2005, pp. 217–228.
2. James A. Bennett, *The Divided Circle: A History of Instruments for Astronomy, Navigation and Surveying*, Phaidon and Christie's Limited, Oxford, 1987.
3. Lucas N. H. Bunt, Phillip S. Jones, and Jack D. Bedient, *The Historical Roots of Elementary Mathematics*, Prentice-Hall, Englewood Cliffs, NJ, 1976.
4. *Catalogue of Keuffel & Esser Co.*, New York, 1909.
5. Allan Chapman, *Dividing the Circle: The Development of Critical Angular Measurement in Astronomy*, 2nd ed., Praxis Publishing, 1995.
6. Robert Coleman, Jr., *The Development of Informal Geometry*, Columbia University Teachers College Bureau of Publications, New York, 1942.
7. Jerry J. Cummins, Margaret Kenney, and Timothy D. Kanold, *Informal Geometry*, Merrill Publishing, Columbus, OH, 1988.
8. Maurice Daumas, *Scientific Instruments of the Seventeenth and Eighteenth Centuries and Their Makers*, trans. and ed. Mary Holbrook, B. T. Batsford, London, 1972.
9. Fisher Scientific Company, Education Catalog, www.fishersci.com, 800-640-0640.
10. Flightstore: The Flight and Model Store, www.flightstore.co.uk.
11. J. L. Heilbron, *Geometry Civilized: History, Culture, and Technique*, 2d ed., Clarendon Press, Oxford, 2000.
12. Alan R. Hoffer, *Geometry: A Model of the Universe*, Addison-Wesley, Menlo Park, CA, 1979.
13. William D. Johnstone, *For Good Measure: The Most Complete Guide to Weights and Measures and Their Metric Equivalents*, 2nd ed.. NTC Publishing, Lincolnwood, IL, 1998.
14. Peggy Aldrich Kidwell, James Prentice's Rectangular Protractor, *Rittenhouse*, 1, no. 3 (1986) 61–63.
15. Peggy Aldrich Kidwell, curator, Slates, Slide Rules, and Software: Teaching Math in America, Smithsonian's National Museum of American History, americanhistory.si.edu/teachingmath.
16. The Mathematics Forum at Drexel, Ask Dr. Math: FAQ, www.mathforum.org/dr.math/faq/faq.impossible.construct.html.
17. William Y. McAllister, *A Priced and Illustrated Catalogue of Mathematical Instruments*, Philadelphia, 1867.
18. Museum of the History of Science, Oxford, www.mhs.ox.ac.uk.
19. *Principles and Standards for School Mathematics*, National Council of Teachers of Mathematics, 2000. Electronic version: standards.nctm.org/document/.
20. School Mathematics Study Group, *Mathematics for the Elementary School: Grade 5 Teacher's Commentary*, Part 2, Yale University Press, New Haven, 1962.
21. James E. Schultz, et al., *Geometry*, Holt, Rinehart and Winston, Austin, 2001.
22. Michael Serra, *Discovering Geometry: An Inductive Approach*, Key Curriculum Press, Emeryville, CA, 1997.
23. William Ford Stanley, *Mathematical Drawing and Measuring Instruments*, 6th ed., E. & F. N. Spon, New York, 1888.

Was Pythagoras Chinese?

David E. Zitarelli
Temple University

Introduction

This article presents two self-contained proofs of the Pythagorean Theorem that are strictly geometric, involving neither measurements nor numbers. The first might have been discovered by Pythagoras in the sixth century BC. The second is due to Liu Hui from about 300 AD. The two proofs show how mathematicians in two ancient civilizations—one in the West (ancient Greece) and the other in the East (ancient China)—deduced a result about right triangles from strictly geometric arguments. We also briefly contrast the geometric approaches with an arithmetic method employed by mathematicians from a third great ancient civilization—the Babylonians. The question posed in the title of this article is borrowed freely from a book by Frank J. Swetz and T. I. Kao [7]. Our purpose here is to show how the radically different civilizations in China and Greece regarded right triangles in a remarkably similar way.

The material in this article is appropriate for students taking geometry for the first time in high school (or perhaps earlier); we provide suggestions for using cut-outs to help visualize the process. The only notion that is assumed is the concept of *congruence*, yet even here it is used in the intuitive sense of placing one figure precisely on top of another. The greatest benefit for beginning students might be an understanding of the nature of *mathematical proof*, because the historical approach adopted here illustrates a type of intuitive argument (based on obvious properties of figures) that preceded formal chains of reasoning that characterize deduction. To better appreciate the two geometric proofs, we describe a tactile approach using congruent triangles (formed by cutting two pieces of a rectangular sheet of paper along a diagonal) and various squares.

Figure 1 shows the familiar 3-4-5 right triangle. The Pythagorean Theorem asserts that the square *on* 5 is equal to the square *on* 3 plus the square *on* 4. Sometimes today we think of squaring as an arithmetic operation, for instance, when writing the Pythagorean Theorem as the square *of* 5 is equal to the square *of* 3 plus the square *of* 4. The distinction between the italicized words *on* and *of* is important. The former reflects a fundamentally geometric idea expressing a relationship among squares constructed on the sides of a right triangle. The latter is an arithmetic expression that harkens back 4000 years to Babylon, whose mathematically adroit scientists uncovered numerous arithmetic properties of right triangles. The most important one determined those triples of integers that could represent the lengths of the legs of right triangles. However, we are not concerned with such arithmetical relationships here.

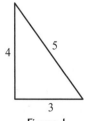

Figure 1

Right triangles in Ancient Greece

Jacob Bronowski (1908–1974) was a scientist of encyclopedic learning who found both the arts and the sciences interesting and accessible. He also had an innate ability to explain complex principles to a general audience. So in 1969, when the BBC sought a scientific counterpart to their highly successful series on Western art called *Civilisation*, they enlisted Bronowski as host and writer. The result was a 13-part series called *The Ascent of Man* that presented the history of science within a cultural history of mankind. The series was an overwhelming success when it appeared on PBS stations in the U.S. in 1973. *Ascent* turned out to be Bronowski's last project; it was completed shortly before his death in New York from a heart attack at age 66. (An informative web site with a short biography of Bronowski, containing extracts from his writings and personal reminiscences from those who knew him, is www.drbronowski.com/.)

The book *Ascent of Man* [1] was published when the show was televised. Chapter 5—correspondingly Episode 5 in the TV series—is devoted to the basic philosophy and lifestyle of the Pythagoreans. Titled "The Music of the Spheres," it emphasizes the Pythagorean credo that mathematical relationships form the essence of all phenomena, and that such relationships alone should be used to express seemingly unrelated events. This particular episode describes the manner in which the Pythagoreans reduced both music and the motions of planets to simple relationships among numbers. It was not until the time of Galileo in the early seventeenth century that scientists again began to look for the essence of things in number. Today this philosophy applies to almost all physical sciences and many social sciences as well. Yet, as we will see, the Pythagoreans could also demonstrate results in a totally geometric manner. The practice of justifying statements on geometric reasoning alone continued through the end of the seventeenth century, when Isaac Newton employed strictly geometric proofs of all results in his historic discovery of calculus.

This section examines a possible proof of the Pythagorean Theorem. By "proof" we mean a convincing demonstration of the correctness of a result. The proof here is a reconstruction due to Jacob Bronowski, who based it on the geology of the region in southern Italy where the Pythagoreans settled. The method of proof lies somewhere between the experimental approach of the ancient Egyptians (from about 1650 BC) and the purely deductive scheme adopted by Euclid in ancient Greece (about 300 BC). We highly recommend that the reader view "The Music of the Spheres" to see how natural the proof of the Pythagorean Theorem might have appeared in Croton, the Italian seaport ultimately inhabited by the merry band of philosophers called the Pythagoreans. Nature presents us with many shapes, like the tiles that dot the Mediterranean coastline. Such tiles often appear in the shape of triangles. (See Figure 67 in [1].)

We describe how Pythagoras's line of reasoning might have been based on tiles to justify the result now named after him. Begin with the right triangle shown in Figure 2. For classroom demonstrations we suggest that the teacher bisect two 8½×11 pieces of paper to form four triangles. We recommend that students bisect 3×5 cards.

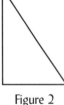

Figure 2

Take a second triangle (a congruent copy of the one in Figure 2), rotate it 90° clockwise, and place it as shown in Figure 3. An important part of Figure 3 is the dotted line segment on the longer side of the second (bottom) triangle. This segment is the difference between the longer side and the shorter side of the triangle. Equivalently, a line segment that is composed of the shorter side plus the dotted line segment is congruent to the longer side.

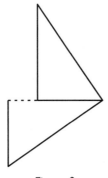

Figure 3

Next, take a third (congruent) triangle and rotate it 180° clockwise from the initial position. Finally, take a fourth triangle and rotate it 270° clockwise from the initial position. Place the four triangles as shown in Figure 4; we number them in their order of construction. If you were to take the initial triangle and rotate it 360° clockwise it would end up where it began. That is the crucial property of a 90°-rotation: after four repetitions, any figure, not just a triangle, will end up exactly where it began.

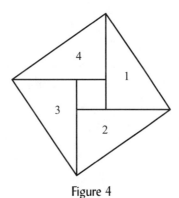

Figure 4

The diagram in Figure 4 is a square (we will call it the *outer square*) composed of four congruent triangles (think of them as four identical tiles) and the *inner square*. Notice that each side of the outer square is the hypotenuse of the right triangle, so its area is equal to the area of a square on the hypotenuse of the right triangle. At this point it is advisable to construct a replica of the inner square; otherwise it might get lost. Each side of this square is the difference between the longer and shorter legs of the triangle; it can be constructed using measurements, if necessary.

The outer square is composed of the four triangles plus the inner square. We aim to rearrange these five tiles—the inner square tile and the four triangular tiles—so they form two square tiles. Physically it is rather easy to slide a tile from one location to another. Begin by sliding tile 1 under tile 3 to form a rectangle; then slide tile 4 under tile 2 to form another rectangle. This results in the L-shaped figure (called a *gnomon*) shown in Figure 5, which has the same area as the outer square in Figure 4 since it is composed of the same tiles.

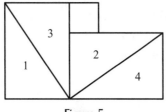

Figure 5

The final step is to draw the vertical dotted line shown in Figure 6 to help visualize how the L-shaped figure can be partitioned into two squares. Look at the figure situated to the right of the dotted line. Each vertical side is the shorter side of the right triangle. Each horizontal side is the difference between the longer side of the right triangle and the side of the inner square, so each one is the shorter side of the triangle. Also, all angles at the four vertices are right, so the figure is a square on the shorter side of the right triangle. Similarly the figure to the left of the dotted line is a square on the longer side of the right triangle. Consequently, the diagram in Figure 6 has been partitioned into two squares, so its area is the sum of the area of the square on the shorter side plus the area of the square on the longer side. Since Figure 6 was formed from Figure 4 by moving tiles whose areas remain the same, the areas of the diagrams in Figure 4 and Figure 6 are the same. Therefore the area of the square on the hypotenuse of the original triangle (Figure 4) is equal to the sum of the area of the square on the shorter side plus the area of the square on the longer side (Figure 6).

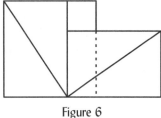

Figure 6

This completes the proof of the Pythagorean Theorem—that's all there is to it. Because there is no extant documentation from the sixth century BC when Pythagoras lived, we have no idea how his followers proved his eponymous result. The proof given here was suggested by Jacob Bronowski 30 years ago.

Bronowski's tactile reconstruction of Pythagoras's proof using tiles does not involve any calculations. No arithmetic was performed—the method is entirely geometric. This approach shows how initially the Pythagorean Theorem dealt with geometric squares and not with arithmetic operations. Next we examine a similar approach that was adopted in an entirely different part of the world—China.

Right triangles in Ancient China

Euclid's famous work on Greek geometry, the *Elements*, was composed about 300 BC, some 200 years after the death of Pythagoras. By contrast, and based on written records, Chinese mathematics is not as old as its Greek counterpart. By the time of the Han dynasty (208 BC – 8 AD), however, there were several Chinese works that reflected a well-developed subject area. One of the major books from this period, *Jiuzhang suanshu*, has been described as the "classic of classics." The title can be translated into English as *Nine Chapters on the Art of Mathematics*, which we shorten to *Nine Chapters*. Just as the *Elements* reflected developments in mathematics since the time of Pythagoras, *Nine Chapters* is a compilation of works written before it. Also like the *Elements*, subsequent commentaries on *Nine Chapters* provided a major outlet for further advances in the field.

Here we examine an impressive, convincing, colorful proof of the Pythagorean Theorem taken from a commentary on *Nine Chapters* by the Chinese mathematician Liu Hui about 300 AD. We will see that Liu Hui's proof is very similar to Bronowski's reconstruction of the Pythagorean proof. The close connection spawned the title of this article. We begin with the same right triangle as Figure 2 from the Pythagorean approach. Here the first step is to construct an inner square on the shorter side in the manner shown in Figure 7. The Chinese colored the interior of this square red; we use dark gray.

Next, construct an exterior square on the longer side of the original right triangle as shown in Figure 8. The Chinese colored the interior of this square blue; we use light gray.

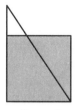
Figure 7

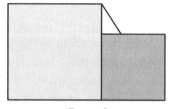
Figure 8

Now construct a congruent copy of the original right triangle. Situate it with the same orientation so that its longer side overlaps the lower-left corner of the light gray square by a triangle congruent to the unshaded triangle in Figure 8. This means that triangles A and B in Figure 9 are congruent, which enables us to transfer the light gray color from A to B. Chinese mathematicians referred to the process of transferring colors as "out-in mutual patching."

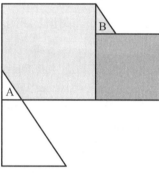
Figure 9

Next, make a copy C of the original right triangle, rotate it by 90° clockwise, and place it into the position shown in Figure 10. Then remove the light gray color from another copy D of the triangle onto triangle C; the transfer of this color is also shown in Figure 10.

So far no dark gray color has been transferred. Can you anticipate the final step involving such a transfer? Look at triangles E and F in Figure 11. They are congruent right triangles with one leg equal to the shorter side of the original right triangle, and the other leg equal to the difference between the shorter side and a leg of the congruent triangles A and B. Therefore the final step is to transfer the dark gray color from triangle E to triangle F.

The shaded region in Figure 11 (consisting of both light gray and dark gray regions) forms a square on the hypotenuse of the original right triangle. The light gray part of Figure 11 initially formed a square on the longer side of the triangle, while the dark gray part formed a square on the shorter side. Because the area of the region that is shaded equals the area of the light gray region plus the area of the dark gray region, it follows that the area of the square on the hypotenuse is equal to the sum of the areas of the squares on the shorter side and the longer side.

This completes the colorful Chinese proof of the Pythagorean Theorem. It is important to point out that Liu Hui's proof, like Pythagoras's, did not use calculations; both are entirely geometric, with no arithmetic computations performed.

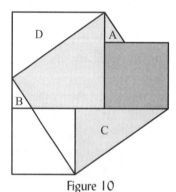
Figure 10

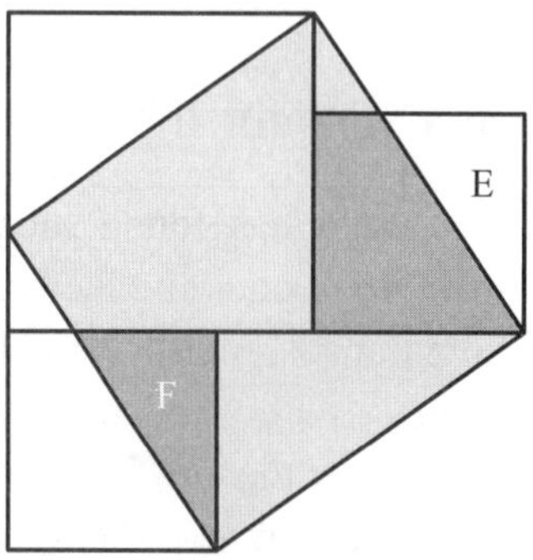

Figure 11

Was Pythagoras Chinese?

It is especially difficult to answer questions about Chinese history during the time that *Jiuzhang suanshu* was written because in 213 BC the emperor Shih Huang-ti ordered all books to be burned. (See [3] and [5] for excellent accounts of Chinese mathematics.) The history of the West makes the question of priority of results and independence of proofs troublesome for the Greeks too. (The book by Heath [4] remains an excellent source for this history.) However, the proofs of the Pythagorean Theorem by Pythagoras (as reconstructed) and by Liu Hui are typical of the types of justification that convinced others of the correctness of geometrical statements about 2500 years ago. The nature of the Greek and Chinese proofs is almost identical. (The textbook [8] provides a similar development.) This raises three questions:

1. Is it possible that one civilization borrowed from the other?
2. Might Pythagoras have preceded Marco Polo in his travels to the Orient?
3. Might an unknown Chinese businessman, schooled in mathematics, have traveled to the Mediterranean area and come in contact with Pythagoreans?

Before asking another question, we hasten to add that there is not a shred of evidence to suggest answers to these three questions. Lacking historical evidence, we can only conclude that Greek and Chinese mathematicians proceeded independently. Now we pose a question for anyone familiar with Euclid's proof of the Pythagorean Theorem (Book 1, Proposition 47).

4. Which proof is more convincing—the one by Euclid or those due to Pythagoras and Liu Hui?

We expect most people familiar with Euclid's proof to answer Question 4 with "those due to Pythagoras and Liu Hui" because they are immediately convincing. This suggests that school districts might improve geometric instruction if a first approach is along historical lines, perhaps enhanced by an excursion into the three-dimensional analog. (For further details, please see "When is a Square Square and a Cube Cubical" by Amy Shell-Gellasch in this volume.) However, an historical approach must contain a vital caveat for anyone desiring to return to the glory days of these ancient civilizations: depending entirely on figures and drawing conclusions from experience can lead to incorrect results. Deduction arose as a response to this shortcoming, and it has served as the mathematical standard since the time of Euclid. (Section 3-6 in [2] discusses how the method of deduction arose, with ties to the ancient Egyptian culture.) Even deduction took a drastic hit by the early 1930s when two logicians, Emil Post and Kurt Gödel, proved independently that in any system like the one needed to demonstrate the Pythagorean Theorem, there will be some statements that can be neither proved nor disproved.

We end this presentation with a question for beginning students.

5. Which proof do you find more convincing—the Chinese or the Greek?

Directions for the instructor

The author has found success with students repeating the Greek proof of the Pythagorean Theorem. However, replicating the Chinese method has been more elusive, and generally results in a demonstration using transparencies. Nonetheless, in this final section we provide instructions for students to perform both proofs.

For the Greek approach, each student will need four congruent right triangles, and a square whose side is the difference between the longer side and the shorter side of the triangle. We instruct each student to cut two 3×5 cards in half to form the four triangles, and then to construct the inner square from a third 3×5 card. The resulting "tiles" are small enough to be manipulated on the student's desk top, yet big enough to be convincing.

Initially, all students should place the four triangles and square as shown in Figure 4. It is important for students to see that each side of the outer square is the hypotenuse of the original triangle. Ask, "How can these five tiles be rearranged to form two squares?" The object is for the students to slide the tiles to the positions shown in Figure 5, and to realize how two squares result. This realization can be confirmed by constructing squares of appropriate size and placing them on top of the gnomon in Figure 6.

Having students reconstruct the Chinese method is not as easy. We think it is best to use graph paper and color appropriate regions, an approach that resonates with the Chinese method. However, we have constructed a method using squares and triangles that can be cut from a sheet of paper with inches as the basic unit. First cut a 4×4 square (denoted S4) and use a red pencil to color it. Next, cut a 3×3 square (denoted S3) and use a regular lead pencil to shade it. Then cut six 3-4-5 right triangles, labeled T0, T1, T2, T3, T4, and T5, with T3 colored red. The proof follows from these five steps.

1. Place S3 atop T0 and S4 to the left of T0 (Figure A).

2. Place T1 at the top left corner of S4 and T2 directly below it (Figure B).

3. Place the red T3 under S3 as shown in Figure C. Then transfer the red from triangle A to triangle B by coloring B.

4. Place T4 atop S3 and to the right of T0. Color the triangular part of T4 that covers S3 with the lead pencil (Figure D).

5. Move T4 into the position in Figure E so its colored region fills the gap, and then place T5 where T4 had been to complete the transfer of color.

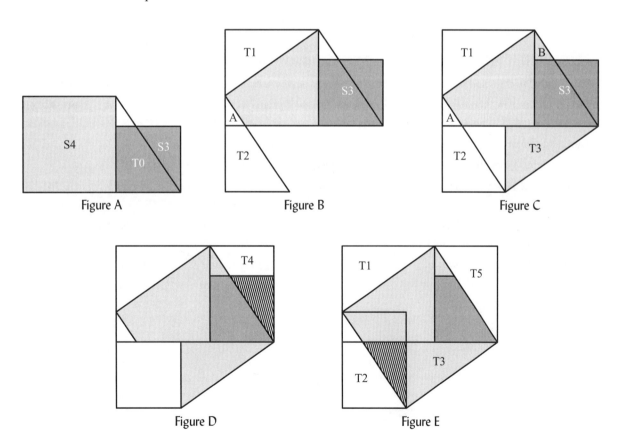

Beyond Pythagoras and Liu Hui

Euclid's book, generally referred to as the *Elements*, consists of thirteen chapters that were then called "books." Major results were called "propositions." Book I consists of 48 propositions, and the penultimate one is the Pythagorean Theorem, labeled merely Proposition I-47. It reads:

In right-angled triangles the square on the side subtending the right angle is equal to the squares on the sides containing the right angle.

Notice that Euclid uses the expression "square on the side" instead of "square of the side." This subtle difference reinforces the geometric nature of the Pythagorean Theorem, as opposed to the arithmetic character suggested by the alternate wording. Yet Euclid went beyond Pythagoras. Proposition 31 in Book VI reads:

In right-angled triangles the figure on the side subtending the right angle is equal to the similar and similarly described figures on the sides containing the right angle.

This result is an impressive generalization of the Pythagorean Theorem. It says, for instance, that the area of an equilateral triangle placed on the hypotenuse of a right triangle is equal to the sum of the areas of equilateral triangles placed on the two legs. That extension would take care of three-sided figures on the legs. The Pythagorean Theorem itself deals with four-sided figures.

Proposition VI-31 goes beyond quadrilaterals, and includes regular pentagons placed on the three sides of a right triangle, regular hexagons, etc. In short, Euclid's generalization gives assurance that the area of a regular polygon on the hypotenuse is equal to the sum of the areas of regular polygons on the legs. And one need not restrict attention to polygons—Proposition VI-31 includes such figures as semicircles as well. However, an examination of additional extensions would take us too far afield. For more details, the interested reader is referred to a recent, charming paper [6].

References

1. Jacob Bronowski, *The Ascent of Man*, Little, Brown, and Co, Boston, 1973.
2. Raymond Coughlin and David E. Zitarelli, *The Ascent of Mathematics*, McGraw-Hill, New York, 1984.
3. Joseph W. Dauben, Ancient Chinese mathematics: The *Jiu Zhang Suan Shu* vs. Euclid's *Elements*. Aspects of proof and the linguistic limits of knowledge, *International Journal of Engineering Science* **36** (1998) 1339–1359.
4. Thomas L. Heath, *The Thirteen Books of Euclid's Elements*, Dover, New York, 1956.
5. Jean-Claude Martzloff, *A History of Chinese Mathematics*, Springer Verlag, New York, 1997.
6. John Putz and Timothy Sipka, On generalizing the Pythagorean Theorem, *College Mathematics Journal* **34** (2003), 291–295.
7. Frank J. Swetz and T. I. Kao, *Was Pythagoras Chinese? An Examination of Right Triangle Theory in Ancient China*, The Pennsylvania State University, University Park/London, 1977.
8. David E. Zitarelli, *A Collaborative Approach to Mathematics*, Condor Book Co., Elkins Park, PA, 1999.

Geometric String Models of Descriptive Geometry

Amy Shell-Gellasch
Pacific Lutheran University
formerly of the United States Military Academy

Bill Acheson
United States Army

Introduction

Many art galleries exhibit sculptures constructed of taut strings or wires strung on wood or metal frames. The genesis of much of this form of art is the static string models originally devised and constructed by Gaspard Monge in the late eighteenth century, and the subsequent articulated models of his student Theodore Olivier in the nineteenth century. These models were constructed as three-dimensional aids in the teaching of descriptive geometry in the nineteenth century.

Simple models that exhibit surfaces such as hyperboloids and warped planes can be constructed for classroom use by the instructor or teams of students. These models can then be used to explore families of surfaces as well as aspects of nineteenth century mathematics and education.

History

With the advent of computer aided design, several subjects have disappeared from school and college curricula. Those related to this chapter are drafting, scientific drawing (as well as drawing in its own right) and descriptive geometry. Computers bring new and wondrous worlds to the classroom and allow us to explore ideas at deeper levels than before. However, there is something to be said for getting your hands dirty, so to speak, and experiencing the physical side of mathematics and science.

Throughout the nineteenth century and into the twentieth, students at the United States Military Academy at West Point, New York, were required to study all of the above mentioned subjects. A visit to the West Point Museum or Special Collections at the Cadet Library reveals amazingly precise and beautiful student drawings, landscapes as well as scientific renderings, by former West Point students such as Ulysses S. Grant. During most of West Point's history[1], engineering was the focus of the curriculum. To this end, descriptive geometry was a required course for students after they had taken their two-year core mathematics sequence of algebra, geometry and calculus.

Developed by Gaspard Monge (1746–1818) in the last decades of the eighteenth century,[2] descriptive geometry is the forerunner of modern-day engineering drawing. One of the concepts developed by Monge was that of ruled surfaces. A ruled surface is one that can be described by the movement of a straight line, or ruling. For example, a cylinder can be described as the surface generated when a line (the generator) traces

---

[1] Founded in 1802, The US Military Academy was the first engineering school in the nation.

[2] For an in-depth discussion of Monge, Olivier and the Olivier models, see *The Olivier String Models at West Point* by Amy Shell-Gellasch, [1].

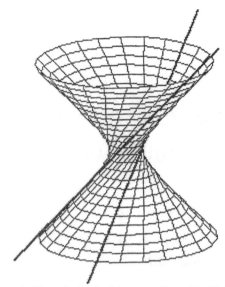

Figure 1. Hyperboloid with two rulings highlighted. [4]

out a circle while remaining perpendicular to the plane of the circle. A set of such lines drawn to depict the surface is called a set of rulings for the surface. A cylinder that is twisted becomes a hyperboloid, also a ruled surface. If a ruled surface, such as a cylinder, can be unrolled to become a flat sheet, then it is called developable. If it can not be unrolled onto a flat plane, it is called a warped surface, such as the hyperboloid in Figure 1.

In order to allow his students to visualize such three-dimensional surfaces, Monge constructed two models by weaving strings onto fixed armatures. Monge's student Theodore Olivier (1793–1853) carried on his mentor's work by writing texts in descriptive geometry that were widely used in universities of the time, both in France and in the United States. Olivier also constructed fifty string models of his own. He had two sets of his models constructed by a Paris firm. One set now resides at the Conservatoire National Musée des Arts et Métiers in Paris, and a little more than half of the second set is now at Union College in Schenectady, New York. The Paris firm that constructed the original models in the 1840s then constructed copies of these models after Olivier's 1853 death. The U.S. Military Academy ordered 26 models in 1857, of which the Department of Mathematical Sciences still has 24.

In contrast to Monge's static models, Olivier's are articulated; the armatures are hinged so that each surface can be manipulated to show a whole family of ruled surfaces. For example, Figure 2 shows a fairly elaborate string model that uses three different colors of strings to display a cylinder surrounding both a circular and elliptical hyperboloid. The top brass disc can be tilted to show various inclinations of the surface. In order to keep the string (rulings) taut, lead weights are suspended from each string and hidden from view inside the base.

A simpler model can be constructed to show only the circular hyperboloid (see Figure 3). On this modern model, the top disc can be twisted in order to depict all hyperboloids from a cylinder to a hyperboloid with a very narrow waist. Detailed steps for constructing this model can be found in the appendix.[3]

In the classroom

Though few if any schools teach descriptive geometry, the history of the subject can be included in many classes besides a history of mathematics course. In addition, constructing a working string model can be

---

[3] For more images of the Olivier models, please visit either the Union College [2] or United States Military Academy Department of Mathematical Sciences [3] websites.

Figure 2. Cylinder and Hyperboloids. First generation Olivier model, c. 1860. Department of Mathematical Sciences, United States Military Academy.

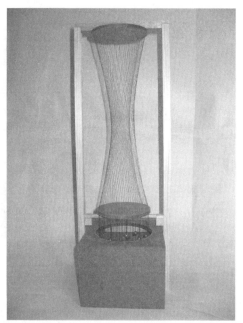

Figure 3. Hyperboloid string model constructed by Bill Acheson, for the Department of Mathematical Sciences, United States Military Academy, 2002.

beneficial, either in concert with the history of descriptive geometry, or on its own. Classes in which this can be incorporated may include calculus, differential equations, geometry, computer design courses, engineering, mechanical arts and, of course, history. For example, in the calculus sequence we often refer to a family of curves or lines. By manipulating a string model, a family of curves can easily be shown, the cylinder/hyperboloid above being a simple one. In a differential equations course, a string model can be used to depict tangents, derivatives and partial derivatives. Since all the Olivier models are of ruled surfaces, this is a topic you may want to add into your geometry or topology courses.[4] The original intention of the models was to help engineering students visualize the intersections of various surfaces; any course that deals with geometric surfaces would be a likely candidate for a unit based on the construction of a string model.

Since the construction of sturdy, aesthetically appealing string models is not a short in-class project, we recommend that the instructor (or a good friend of the instructor who likes to build things) take the time to construct a nice model that can then be used in the classroom year after year. If you have a group of students who are motivated to take the time to build one themselves, encourage them in any way you can. If the hyperboloid is too complex for your course needs, a simpler model could be used. One possibility would be two intersecting cylinders (see Figure 4). This could be a static model for use in calculus courses to show intersection, tangents and volume.

The warped plane is another simple string model (see Figure 5). In this model, a plane can be twisted to form a warped plane. When the top arm is parallel and in the same plane as the station-

Figure 4. Olivier Model of two intersecting cylinders. United States Military Academy Department of Mathematical Sciences.

---

[4] Many college libraries still have descriptive geometry texts, especially from the 19th century, that can be used.

Figure 5. Olivier Model of a warped plane.
United States Military Academy Department of Mathematical Sciences.

ary middle arm and the base, the strings form a plane. Rotating the top armature perpendicular to the plane of the strings results in a warped plane.

For a rough and ready (disposable) string model to be made in class, the following is suggested. For the hyperboloid, make two 12″ discs out of cardboard and punch 10 to 12 holes around the edge. Now thread the same number of pieces of string, each approximately one and a half feet long, from the top disc to the bottom disc, knotting each top and bottom to hold them in place. One person can hold the top disc while another holds the bottom. By rotating one of the discs, the family of hyperboloids can be shown. Four discs and two to four people depending on the size of the discs and the length of string could form the model of two intersecting cylinders.

For the warped plane shown above, two pieces of rectangular cardboard with about twelve 1-inch slits cut perpendicularly along the edge (to look like combs), or even two large plastic hair combs found at the drug store, can be used. Take a long piece of string and weave it up and down from one comb to the other. Again, by having one person holding the top comb and another holding the bottom comb and twisting, the warped plane is depicted.

Conclusion

The string models of descriptive geometry originally made by Monge and perfected by Olivier are beautiful, intriguing, and inspiring mathematical objects. They are part of our mathematical heritage that can be readily shared with students, either directly through the manipulation of models constructed for the classroom, through the actual construction of one by the students, or virtually through the web[5]. Though the construction and use of string models in class requires a fair amount of time and planning, the time spent is well worth the effort. The "lost art" of descriptive geometry is the foundation of our modern engineering drawing, and as such, has a place in both engineering and mathematics education.

---

[5] A wonderful and interactive virtual tour of string models can be found at [2].

References

1. Shell-Gellasch, Amy, "The Olivier Models at West Point", *Rittenhouse Journal of the American Scientific Instrument Enterprise*, **17** (2), December 2003, pp. 71–84.
2. Union College Olivier Models: http://www.union.edu/Academics/Special/Olivier/.
3. United States Military Academy Department of Mathematical Sciences Olives Models: http://www.dean.usma.edu/departments/math/.
4. Wottreng, Kristen, *Computer Methods in Descriptive Geometry and Differential Geometry: Monge's Legacy*, D.A. dissertation, Department of Mathematics, Statistics, and Computer Science, University of Illinois at Chicago, 1999, pp. 111–114.

Appendix

Bill Acheson

This project is a fairly simple reproduction of an Olivier String Model that I first saw while an instructor of mathematics at the United States Military Academy. The basic design can be used for a variety of examples of planar geometry and more or less time can be spent polishing the finished product. I encourage you to experiment and pursue your own modifications. The basic supplies for this project, excluding tools, should cost less than $20.

Naturally, many things in a woodshop can bite. While power tools speed things up and usually make cleaner cuts, they also bite a lot harder when things go wrong. Always know where your fingers are relative to a blade or bit, be absolutely sure to clamp your work down if you use the circle cutter, and wear eye protection. If you are unfamiliar with the equipment, get some help.

Material List
2'×2' sheet of ¼" hardboard (Masonite)
5 lineal feet of ½" square molding
1 lineal foot of ¼" square molding
Wood glue
4 ea. #8 1" screws
1 ea #8 5/8" screw
1 ea #8 finish washer
Small spool of high quality, brightly colored thread
1 sewing needle
60 lead weights — (I used #5 split shot used for fishing)
Masking tape

Tool List (painful minimum)
Handsaw
Electric drill
1/16" drill bit + miscellaneous larger bits
Compass or dividers
Block plane

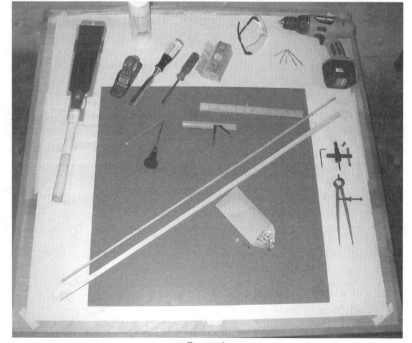

Figure 1.

Square
Scissors
Sharp pencil
Ruler
Screwdriver

Additional Tools for an easier life
Drill press and circle cutter
Table saw
Small clamps

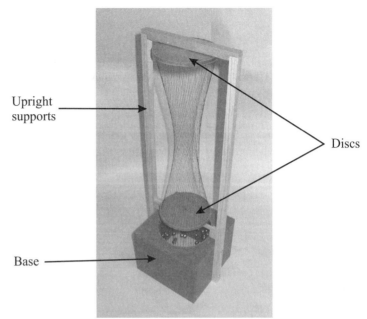

Figure 2.

The model consists of the three basic components seen in Figure 2: the base, the discs, and the upright supports.

The base

To build the base, I first cut two 4″ wide by 24″ long pieces from my sheet of ¼″ hardboard. If you are using a handsaw, cut a hair wide of your line and then plane the cut off smooth to the line. From these 4″ wide panels, I cut 2 smaller panels 5 ½″ long and 2 panels at 5″ long. They will form the sides of your base. The longer 5 ½″ panels used as front and back of the base overlap the 5″ side panels. Keep the remaining 4″ wide stock to cut your discs from later.

I use internal glue blocks cut from the ½″ square stock to reinforce the otherwise very weak butt joints in the case. The length of the glue blocks is not critical as long as when glued up the ends do not extend past the edge of the panel. As seen in Figure 3, I glued the sides of the base in two stages. First, glue the reinforcing blocks flush with the edges of the 5″ wide panels and clamp them using some convenient textbooks as weights. Pay attention to getting the long edges of the glue blocks and panel flush, as doing so will make life a lot easier in the next step when you attach the sides and work to get things square.

Once the glue dries, you will glue the front and back to the sides. Masking tape works well as a clamp while the glue is drying. Check for square. Though it takes some extra time to do this in two steps, imagine trying to hold 8 parts together and square while the glue is drying!

For the top of your base, cut a piece from your remaining hardboard to meet the finished dimen-

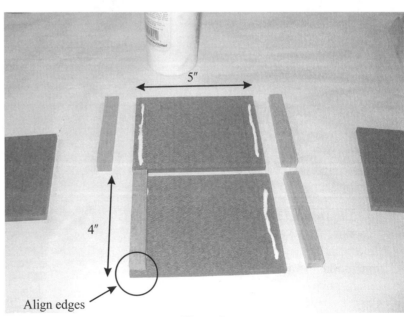

Figure 3.

Geometric String Models of Descriptive Geometry

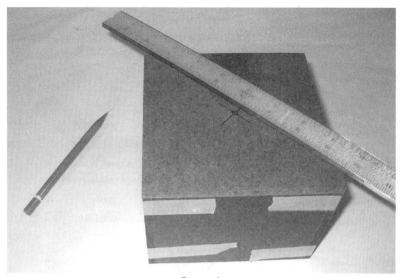

Figure 4.

sions of your base, which should be very close to 5 ½″ square. While the top can overhang front to back, any overhang on the sides will interfere with the uprights. I consider the front and back to be the 5 ½″ wide pieces so I won't have to look at the butt joint when the project is finished. Before setting the base aside for a minute, use intersecting diagonals to mark the center of the top for drilling later as seen in Figure 4.

The discs

The dimensions of the base limit the maximum diameter of the discs as does your ability to cut the discs. If at all possible, I highly recommend you cut your discs using a circle cutter on a drill press. Doing so will be far faster and neater than the alternative of using a compass to scribe a circle on the hardboard, using a coping saw to cut close to the line by hand, and sanding to the line. While theoretically possible to do neatly, I'd rather spend my time doing non-linear optimization in my head. Under no circumstances should you try to use a circle cutter in a hand drill; you will hurt yourself. I chose my disc diameter to be 3 ¾″, though +/- ¼″ is not unreasonable. Whatever diameter you choose for your discs, realize that the opening in the top of the base must be slightly larger for the strings to hang down neatly. Figure 5 shows the set-up I used on my drill press to cut the two discs that hold the strings in place. Make sure you clamp things down; otherwise you risk getting hurt from the spinning circle cutter or having the work piece ripped from your hands. I am guilty of knowing from experience and have a pink scar on my left middle

Figure 5.

Figure 6.

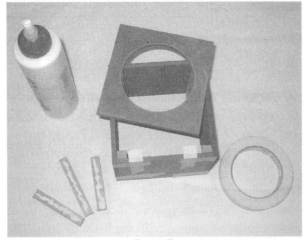

Figure 7.

Figure 8.

finger as testimonial. Another detail is to set the drill press to its lowest RPM setting since the cutter is by design unbalanced and not meant for high RPMs. Don't have a drill press? Try the local high school shop or perhaps your school system has a maintenance shop. Use a sacrifice block of wood underneath your hardboard so that you don't destroy the cutting blade against the cast iron drill press table. After you've cut out your discs, lightly bevel the edges to remove any burr.

Figure 6 shows the circle cutter being readjusted to cut the hole in the top of the base through which all the strings hang down by their weights. Note that I've set the diameter slightly larger than the discs and that I've reversed the cutter tip to the outside as compared to the usual arrangement in Figure 5. This helps cut a clean circle from the inside for the top.

Using the same drill press set-up, cut the circle in the top of the base using the center located by intersecting diagonals. Attach the top to the base, making sure the glue blocks don't interfere with the opening in the top. Again, I use a two-step process for gluing and use masking tape for clamps to hold the glue blocks in place as seen in Figure 7. Before attaching the top, I ease the outside edges to eliminate any rough or sharp spots with my hand plane or sandpaper.

While the glue holding your top in place is drying, you can work on the most tedious part of the whole project. Ultimately to make the project meaningful, you have to get a whole bunch of holes evenly spaced just inside the perimeter of the discs to hold your strings in place. To achieve an even spacing I chose to drill the discs to accommodate 60 strings, or 15 per quarter. Technically that works out to a hole every 6 degrees. Unless you have access to a highly accurate compass, that doesn't do you much good. On a block of scrap, from a center I traced the diameter of the discs in the first quadrant. Using my dividers and an initial guess I marked off segments and adjusted the setting on my dividers until I could divide the quadrant up into 15 sections. Figure 8 illustrates the set up. Only the setting on the dividers is important. While perhaps a little brutish, there is enough natural error in the next step to make any quest for extreme accuracy at this stage to

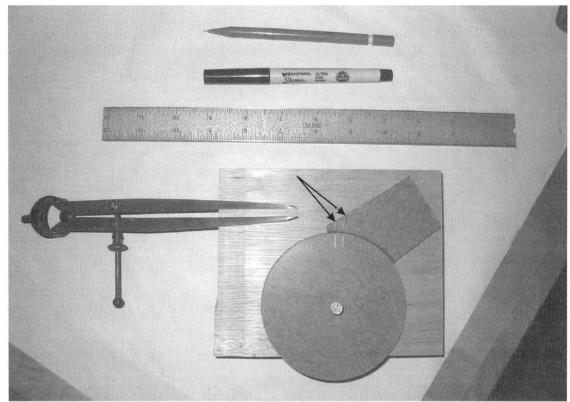
Figure 9.

be useless. (Of course, few strings spaced further apart can be used, though the final product will not look as sleek as the original Olivier models.)

Figure 9 shows the jig I used to drill all the holes in both my discs at the same time. On the same block of wood as in Figure 8, I used a spare wood screw to act as a pivot for the discs which I taped together to ensure a mirror image of each other. Drill both discs at the same time! Otherwise you risk having 60 holes in one disc and 61 holes in the other! (I also know this from painful experience.) I used a piece of disc cutting scrap glued to the block to act as a gauge. From the center I drew an arbitrary radial line across the gauge. Using the spacing painstakingly adjusted on the dividers, I drew a second radial line.

Figure 10 shows the jig in use. Again a drill press makes life much easier, though you could drill your holes using a hand drill. I used a 1/16″ drill bit for my holes and clamped the jig to my drill press so that it would drill a hole about an 1/8″ inside the edge and *be on a radial line drawn from the center to one of my marks on the jig*. The first hole can be drilled anywhere. Subsequent holes are drilled by rotating the previous hole to align with an imaginary radial line drawn from the center to the second mark on the jig. Two pieces of advice: First, don't force the drill. Be patient and expect to have to withdraw the bit from the hole to empty the cuttings at least once, otherwise you will blow out the back of the discs. Second, given the inherent inaccuracy of the jig, space the last two or three holes by eye to eliminate the possibility of a noticeably larger or smaller gap between the first and last holes.

The upright supports

My upright supports are screwed into the sides of the case and stand 14″ above the top of the base. Cut two 18″ long pieces from your ½″ square molding, making sure they are straight and true. Mark center lines on the sides of your base, align an upright over the lines, drill a pilot hole through your upright and into the

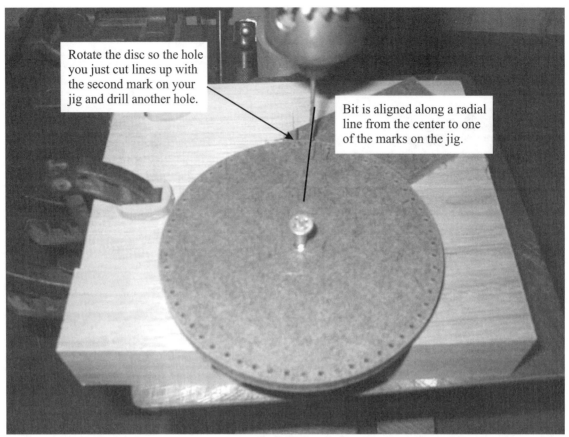

Rotate the disc so the hole you just cut lines up with the second mark on your jig and drill another hole.

Bit is aligned along a radial line from the center to one of the marks on the jig.

Figure 10.

base about ½" from the bottom. Gently screw the upright to the base. Make final adjustments to square the upright to the base and add a second screw. Repeat for the second upright. Figure 11 illustrates the alignment and screw positions.

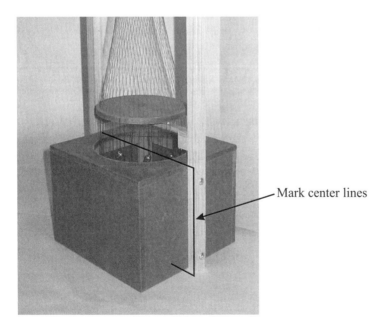

Mark center lines

Figure 11.

The next step is to add the cross beams that support the discs. For the upper cross beam, I cut a piece of ½" stock to length, located the center and drilled a pilot hole as shown in Figure 12. I chose to cut half laps in the ends of my cross beam, but you could just as easily use a butt joint. Figure 12 also shows the finish washer used to space the upper disc away from its cross beam. Gently screw the upper disc to the cross beam ensuring the rough side of the disc is down. Figure 13 shows the cross beam being glued into position using a temporary masking tape clamp.

The lower cross beam is a bit more of a challenge to balance strength with a structure that won't greatly interfere with

Geometric String Models of Descriptive Geometry

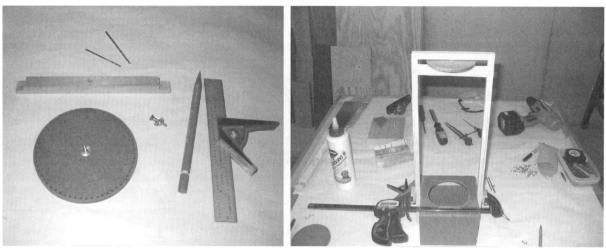

Figure 12.   Figure 13.

the hanging strings. Figure 14 shows a cross beam glued up from pieces of ¼" molding stock. The ¼" stock is thin enough that the strings will hang down past it gracefully, while the added pieces provide strength and surface area to support the disc. If you expect the model to be handled regularly, or by children, I recommend you go for the strength of a cross beam made from ½" square stock instead and accept some disruption of the hanging strings.

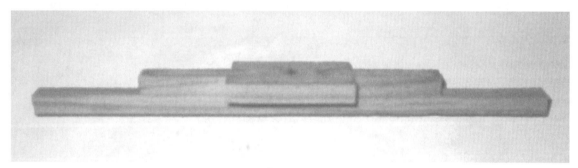

Figure 14.

Before attaching the lower disc to its cross beam, use a few weighted threads to position the disc so that the threads hang evenly inside the hole in the top of the base as seen in Figure 15. I use lead fishing weights to tension the thread. Rather than using a screw, I glued the lower disc to the built up cross beam. As before, make sure the rough side of the disc is down. Figure 16 shows the small blocks of ½" stock I glued to the uprights to support the lower cross beam. For more strength you can glue small pieces of ¼" stock to the shoulders of the joint as indicated.

Once all the glue dries, all that remains is to thread your model. If you want to paint or decorate your model, do it before you thread. To facilitate the threading process I recommend a sewing needle, sharp scissors, and a nail pounded into a piece of scrap wood to hold your spool of thread. I would thread my needle, feed it up through a hole in the lower disc, to a hole in the upper disc and back down through an adjacent set of holes. Figure 17 shows trimming the thread to length and adding the weights. I usually had to tie a knot in the thread to keep the weight from falling off. Stagger the weights as they hang down inside the base to limit their interference with their neighbor. When the upper cross beam blocks your needle, just twist the disc a bit to either side. If you find you have an odd number of holes, tie a thread off using two adjacent top holes and thread it back down the remaining hole.

Figure 15.

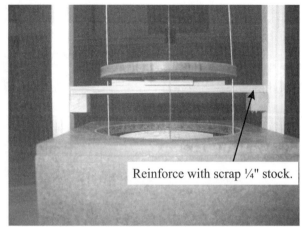

Reinforce with scrap ¼" stock.

Figure 16.

To help keep the hanging thread from tangling inside the base during transport, cut two rectangular pieces of cardboard and tape into cylinders the same height and diameter of the base, one slightly narrower than the other. From beneath, slide the larger cylinder outside the ring of weights and the other inside to keep the weights from swinging too much.

Naturally you can vary the form of your conic by twisting the upper disc, though realize it is only held by friction and tightening up the retaining screw once the model is threaded is a challenge! Congratulations, you are done! Enjoy your own reproduction of a 19th century Olivier String Model (Figure 18).

Figure 17.

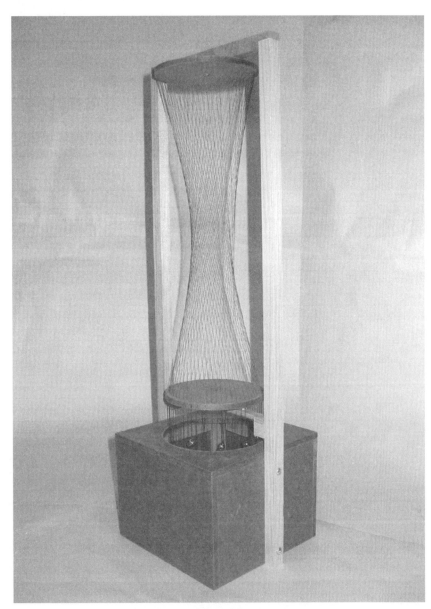

Figure 18.

The French Curve

Brian J. Lunday
formerly of Department of Mathematical Sciences
United States Military Academy

Introduction

Humankind designs and constructs tools, furniture, vehicles, buildings and other structures with deliberate precision. And while it is relatively easy to draw objects using only lines and circles, these shapes are insufficient to represent the constructs of our world. Although it is possible to represent some simple curvatures in two dimensions with a linear projection of an arc or a circle to an ellipse, a parabola, or a hyperbola, these too are insufficient in their lack of complexity. Within these limitations, gross representation of curved lines is not sufficient for engineering purposes. In the words of Professor Thomas French, the writer of the authoritative *Engineering Drawing* series of textbooks in the early 20th Century:

> The engineering draftsman has a greater task [than the artist]. Limited to outline alone, he may not simply suggest his meaning but must give exact and positive information regarding every detail of the machine or structure existing in his imagination. Thus drawing to him is more than pictorial representation; it is a complete graphical language, by whose aid he may describe minutely every operation necessary and may keep a complete record of the work for duplication or repairs. [9, p. 1]

This need to depict and design complex curvatures with accuracy begat the use of the French Curve in the 19th and 20th Centuries.

Applications

The demand for technical drawing actually surged with the Industrial Revolution in the middle of the 19th Century. Prior to this era, the construction of custom-made houses, tools, or other machinery did not require such precision in two-dimensional representation. The Industrial Revolution brought about two trends that required improved precision in drafting: mass-production and the use of interchangeable parts. This need further increased in the early 20th Century as national industrial bases surged to meet the weapons and materiel (referring to military logistics) production requirements of World War I. [3, p. 167]

There are both specific and general drafting design requirements with complex curvatures that necessitate the use of a French Curve. In the general case, the French Curve is useful for drawing any curved line that is not the arc of a circle. Different types of French Curves are helpful in drawing ellipses, parabolas, and hyperbolas. [13, p. 15] As early as 1887, a Massachusetts Institute of Technology textbook prescribed their use for such curves when "a series of points" on the curve were already known. [7, p. 47] Specific applications of the French Curve included their use for designing the curves on ships, [13, p. 14; 10, p. 4] designing the curvature of railroad tracks, [13, p. 15; 10, p. 4] and the construction of steam curves on temperature-entropy diagrams for steam at constant volume, pressure, or percentage in the gas phase. [9, p. 15]

The French Curve

"What exactly is a French Curve?" In the most general sense, a French Curve is a solid, flat template that manifests variable curvatures on its external edges and internal edges for the purpose of drawing a smooth, curve through a series of fixed points. The definition is necessarily broad because the available shapes and materials found among French Curves are diverse. It is difficult to go to a drafting store and buy a single French Curve. They are usually available in sets of three or eight curves of the more commonly used shapes. The actual number of shapes of French Curves, however, is limitless. Although some manufacturers sell up to 22 standard shapes, they can produce large orders of custom-made shapes as well. [11]

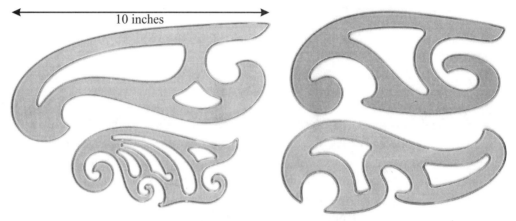

Figure 1. Some common French Curves

The more common designs for French Curves consist of a variety of continuously linked curvatures of a single category of geometric shape. The four most commonly used shapes are ellipses, hyperbolas, parabolas, and spirals. [6] Two of the more common spirals found in French Curve construction are the logarithmic spiral [9, p. 10] and the Spiral of Archimedes. [13, p. 44] In his 1911 text [9, p. 15], Thomas French prescribed the following spiral curve as useful for engineering diagrams and steam curves:

$$r = A\sec(\theta) + k \tag{1}$$

This curve is shown in Figure 2, where $A = 5.5''$ and $k = 8''$ over the following domain:

$$2.2 \leq \theta \leq \pi$$

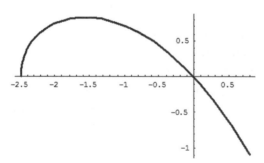

Figure 2. Polar Plot of 1911 French text equation

Materials

The range of materials for French Curves over the last 150 years has varied as much as their shape. As of 1857, they were made of "thin wood, of variable curvature" with no varnish; the wood was kept in its natural state but kept carefully wiped. [16, p. 11] The most commonly used woods were white wood, cherry wood, [12, p. 15] white holly, [2, p. 10] and pear wood. [10, p. 4] However, the industrial revolution brought experimentation with different construction materials. By 1894, French Curves were often made with rubber or celluloid. [2, p. 10] These materials had their disadvantages, and draftsmen either shunned them or adapted techniques to improve their functionality. The rubber was deemed poor for use as a French Curve because "it is difficult to make pencil marks on its surface." [2, p. 10] Celluloid had the same problem, but draftsmen learned to overcome this by sandpapering the surface to roughen it. [2, p. 10] By the time Thomas French wrote *Engineering Drawing*, he deemed celluloid "the only material to be considered" for French Curves, though he acknowledged that home-made curves could easily be constructed out of "thin holly or bass wood, sheet lead, celluloid, or even cardboard or pressboard." [9, pp. 14–15] As plastics technology continued to develop and improve, its inexpensive and durable nature ensured it as the future material for French Curves. Today, French Curves are usually constructed from transparent green, grey, clear, or amber plastic.

Nomenclature

Textbook authors throughout the late 19th century and early 20th century generally attribute the development of French Curves to France, but none pinpoints their origin or their inventor. The French Curve has not always gone by that name, though. Even today, the French call it "le pistolet (á dessin)" [6] which, literally translated, means "the drawing pistol" or "the design pistol." As early as 1857, they were known as "variable curves" or "universal sweeps." [16, p. 10] Heavily influenced by French descriptive geometers like Monge, Dennis Hart Mahan referred to a French Curve as a pistol, pistolet, or a curved ruler in his 1870 text. [12, p. 15] By the late 1800s, the first manuals began calling these tools "French Curves" or "irregular curves." [1, p. 29; 7, p. 9; 5, p. 9] By the turn of the century, some manuals and textbooks had still not accepted a standardized terminology for the instrument. An 1898 Yale University text refers to them as "curve-rulers," [15, p. 29] while a 1905 U.S. Military Academy text refers to them as irregular curves. [10, p. 4]

In the early 20th century, several people sought to write authoritative drafting manuals to encompass and collate current knowledge, standardize practices, and gain prestige for drafting as a technical field. At least two of them had notable success. Thomas Ewing French penned *A Manual for Engineering Drawing* in 1911. In this manual, he used the terms "curved rulers, irregular curves, or French Curves" [9, p. 14] interchangeably. It is mere coincidence that a man named French wrote the book and used the term "French Curves"; Goerge André's 1874 text used the term when Thomas Ewing French was only three years old. In 1918, Carl Lars Svensen penned *Essentials of Drafting* as a competing text. In his book, Svensen refers to French Curves as "irregular curve[s]" and lists them among "the really necessary equipment" for drafting, but never references how to use them in 184 pages of printed material.[14, p. 1] Although French passed away in 1944, a revised and updated version of his work is still available in its 14th edition of 1993 as *Engineering Drawing and Graphic Technology*. Meanwhile, Svensen's original work did not last as long; the 20th and final printing of its 2nd edition occurred in 1947. However, a later collaborative effort of French and Svensen known as *Mechanical Drawing* is still available in its 11th edition of 1990. By the mid 20th century, the term "French Curve" had gained widespread acceptance as the term of choice in other manuals and textbooks as well. Today, the terms "French Curve" and "irregular curve" are often used interchangeably.

French Curves are not the only tools used for drawing non-circular lines, but they are one of the more versatile drafting tools from the 19th and 20th centuries. French Curves are a specific taxonomy within the broader category of "irregular curves." Irregular curves also include splines, ships curves, railroad curves, and flexible curves. A spline is an adjustable curve made of flexible strips of metal, rubber, or wood. Splines usually have a cross-section in the shape of the letter 'H,' so that three to four pound weights — known as 'ducks' — can be clipped to the strips to hold them in the shape of a curve. The use of splines is restricted to long, sweeping curves such as those found in early 20th century aircraft and automotive design. [9, p.

10] Ships' curves are another category of irregular curves. Specific to ship hull design only, these curves are available in sets of 45–120 curves in a professional set. [13, p. 14] Railroad curves are similar to ships curves, with a specific purpose and a limited range of use. The fifth category of irregular curves is flexible curves. These curves usually consist of layers of thin metal strips, coated with rubber or soft plastic. The advantage of flexible curves is that the draftsman may bend them into a fixed position without the use of weights. The disadvantage of flexible curves is that they are only so flexible; if a draftsman bends them too tightly, the metal inside will fatigue and break.

The significance of French Curves

The logical teaching points behind the use of French Curves are simple. Life is nonlinear, both in natural and man-made constructs. It is not neat; it is not succinct; and it can rarely be described accurately with a simple equation. Our efforts to represent existing behavior are merely approximations. So are our designs for new materials and objects. When it comes to curvature, the implicit goal for approximations is increased accuracy. While computer technology does a better job of nonlinear curve-fitting and regression, French Curves were an important tool to communicate complex curvature for over a century. Moreover, computers have not yet completely replaced French Curves. For drafting by hand, French Curves remain an inexpensive method of curve-fitting that maintains precision.

French Curves and the classroom

The primary use for French Curves in today's classroom is to teach about regression, interpolation, and extrapolation in a more conceptual manner. Using technology, it has become easier than ever to conduct a nonlinear regression of data. With readily available commercial spreadsheet software, students can follow a rote procedure to fit exponential, logarithmic, or polynomial equations to a given set of data. Many also understand that a higher coefficient of determination indicates a better "goodness of fit" for the model to their data. Based on this metric, they often will not hesitate to use their model for interpolation and extrapolation. Unfortunately, many students often do not differentiate between the coefficient of determination and the actual suitability of an equation for a model. While a sixth-order polynomial regression will perfectly fit any series of six data points, it is unlikely to be worthwhile as a model to predict behavior at any other point in time. If students conduct graphical regression using a French Curve before analytical regression with computers, we can impart a greater understanding of the trade-offs between how well a model fits data and how well a model may predict unknown values. The students can also gain a greater appreciation for the level of uncertainty when extrapolating predictions from a model as opposed to interpolating intermediary values.

Another use for French Curves in the classroom is to inspire and motivate the general study of polar plots. Spirals offer one manner to pique student curiosity: "What is a Logarithmic Spiral? What is the Spiral of Archimedes? What is the difference? And which is the closest approximation to a given curvature?" Based on the spiral-based French Curve shown in Figure 3, can you tell?

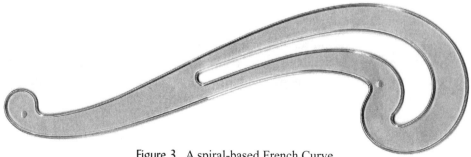

Figure 3. A spiral-based French Curve

A Logarithmic Spiral is a spiral that grows such that the angle between the radial line and the tangent line at any given point is constant. This is also known as the "Equiangular Spiral" and takes the form:

$$r = ae^{b\theta} \qquad (2)$$

where a and b are constants.

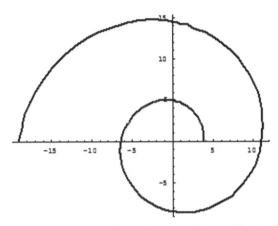

Figure 4. Logarithmic Spiral where $a = 3.75$ and $b = 0.1725$

Meanwhile, a Spiral of Archimedes takes on the form $r = a\theta$, where a is a constant. Its shape is noteworthy due to the equal distance between concentric rings along any radial line.

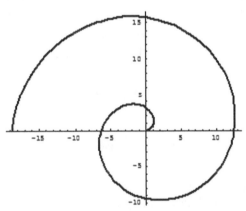

Figure 5. Spiral of Archimedes where $a = 2$

Within the topic of polar plots, French Curves can also motivate the study of conic sections. The three most commonly sold French Curves are based on conjoining subsections of various parabolic, hyperbolic, and elliptical curvatures. Further examples of these French Curves are shown in Figure 6.

While the shapes alone may be useful to inspire conic section studies, it is also useful to conjoin this academic endeavor with polar plotting. Students can work with the following equation [4]:

$$r = \frac{p}{1 + e\cos(\theta)} \qquad (3)$$

Figure 6. French Curves constructed with elliptical, hyperbolic, and parabolic sections

where p and e are constants, and e represents the eccentricity of the function (not Euler's constant). Of note, when $e = 0$, a circle results. Values between 0 and 1 for e produce ellipses; a value of $e = 1$ creates a parabola; and a hyperbola results from values of e that exceed 1. [4] While Equation 3 can be expressed in many equivalent forms, this example is a simple representation.

One more use for the French Curve is as an icon of history and technological development. Almost any physical construction has improved with the tools for design and construction. Aircraft are an excellent example. Students can identify an airplane from each decade of the 20th Century that demonstrates cutting edge technology. From the biplanes of World War I through the stealth aircraft of the 1990s, the lines of each design have become smoother and more deliberate. Within these examples, students can extrapolate the tools used to design the aircraft. The early 20th Century airplanes relied on NACA wing sections and relatively straight lines for body design. As French curve use declined in favor of early computer design, some lines became straighter. The F-117A Nighthawk (stealth fighter) has very few rounded edges because the computer technology wasn't sufficiently mature when it was designed in the 1970s. Within the next decade, advances in computers allowed the development of the smooth bodied B-2 Spirit (stealth bomber), which manifests almost no sharp edges. Other examples in the evolution of design are also available for study, and it can be fun for students to try to identify the tools used to create them.

A final way that French Curves have already been used in the modern mathematics classroom is anecdotal and humorous, but does demonstrate an interesting evaluation of student's conceptual understanding of calculus. From his book, *"Surely you're joking, Mr. Feynman": Adventures of a Curious Character*, physicist Dr. Richard Feynman relates the following story from his days as a teaching assistant.

> I often liked to play tricks on people when I was at MIT. One time, in mechanical drawing class, some joker picked up a French curve (a piece of plastic for drawing smooth curves — a curly, funny-looking thing) and said, "I wonder if the curves on this thing have a special formula?"
>
> I thought for a moment and said, "Sure they do. The curves are very special curves. Lemme show ya," and I picked up my French curve and began to turn it slowly. "The French curve is made so that at the lowest point on each curve, no matter how you turn it, the tangent is horizontal."
>
> All the guys in the class were holding their French curve up at different angles, holding their pencil up to it at the lowest point and laying it along, and discovering that, sure enough, the tangent is horizontal. They were all excited by this "discovery"— even though they had already gone through a certain amount of calculus and had already "learned" that the derivative (tangent) of the minimum (lowest point) of any curve is zero (horizontal). They didn't put two and two together. They didn't even know what they "knew." [8, pp. 36–37]

Conclusion

To average students, a set of French Curves seems like an anachronism. Bring some into your classroom and prove them wrong! French Curves are useful for generating an interest in history, demonstrating mathematical concepts in the classroom, and inculcating a greater, practical understanding of mathematical principles as well.

References

1. André, George G. *The Draughtsman's Handbook of Plan and Map Drawing*. New York: E. and F. N. Spon, 1874, pg. 29.
2. Anthony, G. C. *Elements of Mechanical Drawing: The Use of Instruments; Theory of Projections and its Application to Practice; and Numerous Problems Involving both Theory and Practice*. Boston: D.C. Heath and Co., Publishers, 1904, pg 10.
3. Booker, Peter Jeffrey. *A History of Engineering Drawing*. London: Chatto and Windus, 1963, pg 167.
4. Calvert, J. B. *Parabola* [online article]. Denver, CO: University of Denver, May 3, 2002. Accessed October 16, 2003. Available from http://www.du.edu/~jcalvert/math/parabola.htm; Internet.
5. Cross, Anson K. *Mechanical Drawing: A Manual for Teachers and Students*. Boston: Ginn and Company, 1896, pg. 9.
6. Daube, Klaus. *Drawing and Illustrating in the Pre-digital Time* [online article]. Zurich: Docu+Design Daube, June 5, 2003. Accessed October 8, 2003. Available from http://www.daube.ch/docu/glossary/drawingtools; Internet.
7. Faunce, Linus. *Mechanical Drawing*, 13th Edition. Boston: Linus Faunce, 1887, pp. 9, 47.
8. Feynman, Richard Phillips. *"Surely you're joking, Mr. Feynman": Adventures of a Curious Character*. New York: W.W. Norton, 1985, pp. 36–37.
9. French, Thomas E. *A Manual for Engineering Drawing*, 1st through 6th Editions. New York: McGraw-Hill Book Company Inc., 1911, 1918, 1924, 1929, 1935, 1941, pp. 1–15.
10. Hagadorn, Charles B. *Drawing Instruments and Papers*. West Point, NY: Department of Drawing, USMA, 1905, pg. 4.
11. Huang, Alice. *Discussion by telephone* on June 6, 2003. Hua Ching Manufacturing Company, Republic of China.
12. Mahan, Dennis Hart. *Industrial Drawing*. New York: J. Wiley, 1870, pg. 15.
13. Svensen, Carl Lars. *Drafting for Engineers*. New York: D. Van Nostrand Company, Inc., 1935, pp. 14–15, 44.
14. Svensen, Carl Lars. *Essentials of Drafting*, 1st Edition. New York: D. Van Nostrand Company, Inc., 1918, pg. 1.
15. Tracy, J. C. *Introductory Course in Mechanical Drawing*. New York: Harper Brothers Publishers, 1898, pg. 29.
16. Worthen, W.E., ed. *Appleton's Cyclopaedia of Drawing*. New York: D. Appleton and Company, 1857, pp. 10–11.

Area Without Integration: Make Your Own Planimeter

Robert L. Foote
Wabash College

Ed Sandifer
Western Connecticut State University

Introduction

Clay tablets from Mesopotamia and papyri from Egypt provide evidence that work with area has been part of mathematics since its early history. These Ancients knew how to find areas of squares, circles, triangles, trapezoids, and a number of other shapes for which we no longer have names.

Like many other physical quantities, we usually measure area indirectly. That is, we measure something else, such as lengths, a radius, or angles. Then we do some calculations to find area based on appropriate formulas. The object determines the formula we use and the measurements we make.

There are a number of methods to measure volumes. Some of them are indirect, as for areas, involving linear and angular measurements and the use of formulas. Interestingly, some of them are more direct. One example is the measurement of liquid when cooking, using calibrated measuring cups. Similarly, the volume of a non-porous solid can be measured by submerging it in water to see how much water it displaces (it helps if the solid doesn't float).

Many instruments are cleverly designed devices that convert a value we want to measure into some scale that we can read directly. For example, thermometers (the bulb type, not the digital ones) convert the volume of the liquid (this used to be mercury, but now it is usually tinted alcohol) in the bulb into a length. When the liquid expands or contracts with the temperature, the calibrations along the tube let us read the temperature that corresponds to volume.

Our goal is to learn how to make and use a simple device called a planimeter that measures area more directly without having to measure lengths or angles or to make any calculations. A second goal is to understand the mathematics behind how planimeters work.

Small Project #1. Find out how various instruments convert quantities into measurements that we can read directly. There are thousands of different instruments, but you might start with some of the following:

- Barometer — measures air pressure
- Anemometer — measures wind speed
- Speedometer
- Protractor
- Pitot tube — measures speed of an airplane through the air
- Volt meter
- Bathroom scales

- pH meter
- Gasoline pump
- Tire pressure gauge
- Odometer

There are "digital" models of many instruments available. Usually, the digital models work the same as the analog models, but they have a "digitizer" built in to convert the analog reading into a digital value. This gives the illusion of accuracy and makes it harder to figure out what the instrument is actually doing, so we recommend doing this project with analog instruments rather than digital ones.

This project can be a minor homework assignment by having students give a one- or two-line description of how each instrument works. You can build extra credit into the assignment by having students collect as many different instruments as they can find. It can be a more substantial project if students make posters or web pages showing how the instruments work.

Map readers or map distance measurers

Before we understand area, we should make sure we understand length. We can measure straight lengths directly with a ruler. We can use a tape measure to measure around corners and across certain curved surfaces. We can measure more twisted curves indirectly with a string by running the string along the curve and then measuring the string. We could mark the string in order to use it to make direct measurements.

All of these techniques become more difficult if the length to be measured is longer than the device used to measure it. Nobody would want to measure a distance of a mile (or kilometer) or more with a string or a meter stick. Instead, there is a device called a surveyor's wheel, a wheel of known circumference, usually either a meter or three feet, with a counter attached. The counter measures the distance we roll the wheel by counting the number of times the wheel turns. This is also how the odometer on an automobile and a "cyclometer" for a bicycle work.

Cyclometers and odometers work well for measuring long curves. For measuring short distances, such as a path on a map, there is a device called a map reader or a map measurer. It works similarly to a surveyor's wheel. Figure 1 shows a photograph taken from an eBay auction [4]. Its description follows.

Here is an old instrument used in World War 2 for measuring distances on maps. This precision instrument was manufactured by the Hamilton Watch Company which is one of the oldest and finest watch companies ever. This instrument was manufactured by the Allied Products Division of Hamilton Watch Co. between the years of 1941 and 1957 for the military, then for industrial use afterwards. This one is the "331" model, and is used for distance measurement on maps by using the small roller wheel on a map from point A to point B. This instrument is approximately 5 inches from tip to tip. The gauge is graduated in inches and centimeters.

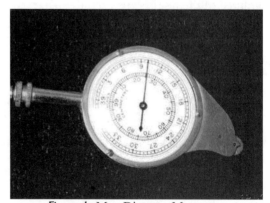

Figure 1. Map Distance Measurer.

These devices are based on a simple idea. You roll the little wheel (on the right side of the picture) along the route on the map. The device mechanically keeps track of how much the wheel turns. This is recorded on a scale, which can then be read. This will be a fundamental idea in understanding how a planimeter works.

Area Without Integration: Make Your Own Planimeter

When using a map reader, it is important to keep the little wheel aligned with the path being measured. Otherwise, the wheel slides sideways a bit and does not measure the full length of the path.

Herein lies a theorem. If the wheel does not stay aligned with the path, then what does it measure? It measures just the component of the path that is parallel to the alignment of the wheel. Here is a more precise statement.

Theorem 1 (Map Reader Theorem) *If a map reader rolls along a curve aligned at a constant angle θ to the direction of the curve, then the distance measured by the map reader equals the length of the curve times $\cos\theta$. In symbols, $d = L\cos\theta$, where d is the distance measured by the map reader (the amount the wheel rolls) and L is the length of the curve.*

It is easy to imagine moving the map reader at a nearly fixed angle to the direction of the curve, due to the way you hold it. It takes practice to get accurate results. Say, for example, that you measure a path that is 20 inches long, and you consistently hold the map reader $7°$ off from the direction of the curve. Instead of reading $d = 20$ in, it would read $d = 20 \times \cos 7° \approx 20 \times .9925 = 19.85$ in.

Of course the angle between the direction of the wheel and the direction of the curve will not generally be constant. In this case it takes calculus to give a precise description of how much the wheel turns—it is $d = \int \cos\theta\, ds$, where θ can now be a variable angle between the directions of the wheel and the curve, and ds represents the length of an infinitesimal piece of the curve. (This just means that the formula $d = L\cos\theta$ is applied to very short parts of the curve and then the results are added up.) Fortunately you do not need calculus to understand the basic idea: if the wheel is not aligned with the curve, the amount the wheel rolls is something less than the length of the curve. The amount that the measurement is off depends on how bad the alignment is.

Here is an example where the angle between the directions of the wheel and path is not constant in which it is pretty clear how far the wheel rolls. In Figure 2, the wheel of a map reader follows the pictured curved path, however, the user keeps the wheel aligned straight "north" on the map instead of aligning it with the path. In this situation, the map reader will always record the distance it is north of the starting point, ignoring the east and west parts of the motion. This is true even if part of the curve has a southerly component, since on this part of the curve the wheel will roll backwards and the distance is subtracted. If we have a coordinate system with the initial point as the origin, then the distance measured by the wheel is the y-coordinate of the final point. Turning the wheel $90°$ to the right would make the map reader measure the x-coordinate or how far east it has moved from the initial point.

One is tempted to say that the map reader measures "positive" distance when it rolls forward and "negative" distance when it rolls backwards, but this can be confusing since distance is inherently positive.

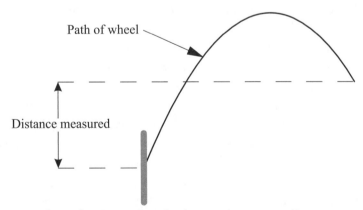

Figure 2. Map reader wheel measuring a y-coordinate.

Nevertheless, this is a useful notion, which we will call "signed" or "directed" distance. Thus the map reader measures signed distance, which is the distance it rolls forwards minus the distance it rolls backwards.

Small Project #2. Make a map reader out of TinkerToys®. The wheel of the map reader can be one of the 2" spools that lets a stick pass through the center without gripping it. In order to count how many times the measuring wheel turns, it helps if you make one or more marks on the circumference. The marks might be every 30° or 45° around the circumference so you can easily see the number of rotations (including fractions) the wheel rolls. The signed distance is then the number of rotations times the circumference.

Use the map reader to measure a variety of distances, lengths of paths on maps, and x- and y-coordinates of points. To begin with, note that the measuring wheel is approximately 2" in diameter, and so about 6.28" in circumference. Start by measuring some line segments of known length. If you get results that are consistently too large or too small, you need to use better value for the circumference. You can get this by rolling the wheel on a piece of paper, marking consecutive places where a point on the circumference contacts the paper, and measuring the distance between these with a ruler.

Lengths and points with coordinates on the order of six to eighteen inches are a good size for this project. This project, especially measuring coordinates along curved paths, should give students a good sense of how the main moving part of a planimeter works. It should take only about 20 minutes.

Measuring area: planimeters

How do you find the area of an irregular region for which there is no formula, such as the area of a lake on a map? You use a planimeter—a mechanical instrument that measures areas. The user traces the boundary of a region with the "tracer point" of the planimeter. Upon returning to the starting point, the area of the region can be read from a dial on the planimeter. We will consider two types of planimeters in this section, polar and linear, and a third type in the last section, the Prytz planimeter. Here are some significant dates in the history of planimeters:

- ca. 1814–1853. Bavarian engineer J. M. Hermann and Italian professor Tito Gonnella independently developed the first planimeters in 1814 and 1824, respectively, but their work was not widely known until much later. (For an animation of a Hermann-style planimeter, see [13].) In 1826 the Swiss engineer Johannes Oppikofer also designed a planimeter. It is unclear whether Oppikofer knew of the earlier inventors' work. Oppikofer's design was the first to be widely known as well as the first to be manufactured and marketed, and thus is considered by Henrici to be the "starting-point" for those that followed. These early planimeters were mechanically complicated, bulky, awkward to use, and it was difficult to get consistently accurate results with them. Other inventors made attempts to address these problems, but were only partially successful [8].

- 1854. Swiss mathematician and inventor Jacob Amsler developed the polar planimeter and a few years later the linear planimeter. They were mechanically simple, small, portable, easy to use, and accurate. They were so clearly superior to existing planimeters that the older planimeters were quickly made obsolete. Since then a few modifications have been made either to improve accuracy or for specific applications [8]. Nearly all modern planimeters use Amsler's basic design, and even those that do not use his design still work on the same mathematical principle. We will see how to make and use a polar planimeter shortly.

- 1875. Danish mathematician and cavalry officer Holger Prytz developed the hatchet planimeter as an economical alternative to Amsler's planimeters. This had many of the advantages of Amsler's planimeter (in fact it was mechanically simpler), but it was less accurate [8, 10, 5]. We will see how to make and use a hatchet planimeter in the last section.

- ca. 1980–present. Digital planimeters. These have digital readouts and built-in calculators, but are mechanically the same as Amsler's planimeters and not noticeably more accurate. They are still manufactured and sold (see some of the links on [6]).

Amsler's Polar and Linear Planimeters

In 1854, Jakob Amsler invented the polar planimeter, a brilliant and simple device for measuring the area of a region. At the time he was still a student at the University of Koenigsberg. He made a career for himself manufacturing tens of thousands of them, and inventing and manufacturing related instruments.

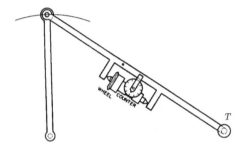

Figure 3. Polar planimeter [9].

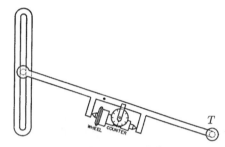

Figure 4. Linear planimeter [9].

Schematic drawings of polar and linear planimeters are shown in Figures 3 and 4. The main part of each is a movable rod, called the tracer arm, with a tracer point at one end (labeled T). A wheel is attached to the rod with its axis parallel to the rod. The wheel is equipped with a scale typically calibrated in square inches or square centimeters. It is similar to a map reader wheel in that it can roll both forwards and backwards, and we will call it the measuring wheel. In a linear planimeter, the end of the tracer arm opposite the tracer point is restricted to follow a linear track, along which it can slide freely. In contrast, in a polar planimeter, the tracer rod is hinged to a second rod, the pole arm, forming an elbow. The end of the pole arm opposite the hinge, called the pole, is fixed so that the pole arm can pivot around it. Consequently the elbow follows the arc of a circle as it moves.

To operate a planimeter, the user selects a starting point on the boundary of the region to be measured, places the tracer point there, and sets the counter on the wheel to zero. The user then moves the tracer point once around the boundary of the region, as shown in Figure 5. The tracer point is typically a stylus or a point marked on a magnifying glass to facilitate the tracing. In a polar planimeter, as the tracer point moves, the elbow at the hinge will flex and the angle between the pole arm and the tracer arm will change. In a linear planimeter, the end of the tracer arm in the track will slide along the track. In both planimeters the wheel rests gently on the paper, partially rolling and partially sliding, depending on how the tracer point is moved. If the pointer is moved parallel to the tracer arm, the wheel slides and does not roll at all. If the pointer is moved perpendicular to the tracer arm, the wheel rolls and does not slide at all. Motion of the pointer in any other direction causes the wheel to both roll and slide. When the tracer point returns to the starting point, the user can read the area from the scale on the wheel.

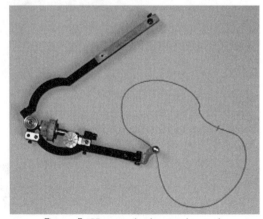

Figure 5. How a planimeter is used.

Like a map reader wheel, the measuring wheel measures a directed distance, namely, the component of its motion that is perpendicular to the tracer arm. The mathematics of how this directed distance can be interpreted as the area of the region traversed by the tracer point is both simple and elegant, and is the basis of how a planimeter works.

Polar and linear planimeters are mechanically simple: polar planimeters need only three moving parts; linear planimeters need only two. Since their invention they have been inexpensive enough that many engineers, architects, surveyors and others could afford to own one. In the electronic age planimeters are less commonly used since the area of a digitized region can be found with an appropriate computer program, but many antique planimeters are still reliable and in service. Modern electronic planimeters are typically expensive, however the electronics are illusory. They are mechanically about the same as their non-electronic ancestors—Amsler would easily recognize them. Furthermore, they are not noticeably more accurate, since their accuracy lies in the skill of the operator, not in the electronics.

Making a polar planimeter

You can easily make a polar planimeter with TinkerToys®. The parts pictured in Figure 6 are from a 2002 Collector's Edition Motorcycle set. It has enough pieces to make three model planimeters.

One polar planimeter uses the following parts (see Figure 6):

4	Spool A	2″ diameter spools that hold sticks tight at the center of the circle
4	Spool C	2″ diameter spools that let sticks pass through the center
2	Long sticks	10.5″ long
2	Short sticks	3″ long
3	Connectors	2″ plastic pieces that connect to spools at the ends and allow sticks to pass through the middle. Some sets do not contain this part.

Assemble it as shown in Figure 7. The figure shows enough parts for *two* planimeters. One set is assembled with the second set placed around it so you can see where each part is used. You can think of the assembly as an arm, with the elbow in the center, the shoulder at the left, and the wrist at the right.

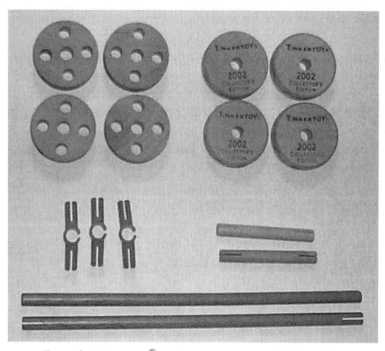

Figure 6. TinkerToys® needed to make a polar planimeter.

Area Without Integration: Make Your Own Planimeter

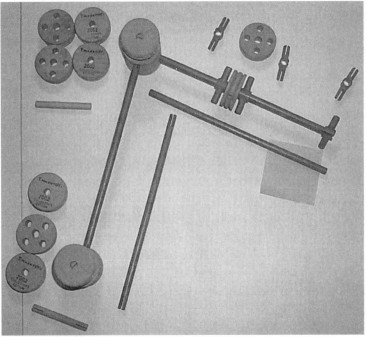

Figure 7. Making a polar planimeter.

The shoulder (which will be the pole) is made of A-Spools connected by a short stick through a C-Spool. The elbow is like the shoulder, but with two C-spools. Join the shoulder to the upper C-Spool in the elbow with a long stick (pole arm). The forearm (tracer arm) is a long stick connected to the lower C-Spool on the elbow. At the center of the forearm, put a C-Spool (measuring wheel), using connectors to keep it as near to the center of the forearm as you can manage, but so that it can still roll smoothly. (If you did Small Project #2 and made marks along the circumference of the wheel, you can use the same wheel for the planimeter. See Small Project #2 for other suggestions.) Align the connectors horizontally so they do not interfere with the rolling. Put the third connector at the wrist, oriented diagonally to the table. This connector is the tracer point. It should be as vertical as possible without reducing the contact of the roller with the surface it rests on. If you like, you can trim the pointer. It should be short enough so it can be oriented vertically, and long enough so it just about touches the surface of the table.

When using the planimeter as described above, you may need to hold the pole to keep it from moving, but be sure to allow the pole arm to pivot around it. After moving the tracer point once around the boundary of the region, as in Figure 8, make a note of the directed distance the measuring wheel rolls. The directed distance is the number of rotations, including fractions, times the circumference. (If you are concerned about the sign of the directed distance, that is, whether it is positive or negative, it depends on which direction you take to be the forward direction of the wheel and which direction you move the tracer point around the boundary of the region. If the net directed distance is negative, you should switch one of these, but not both.)

The following theorem states the amazing fact that makes this a useful device. Here and later it is convenient to abbreviate "the directed distance the measuring wheel rolls" as "the roll of the (measuring) wheel."

Theorem 2 (Planimeter Theorem) *The area of a region measured by a polar or linear planimeter equals the*

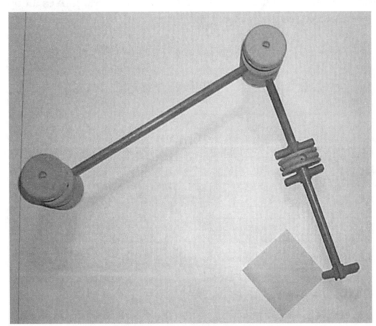

Figure 8. Measuring the area of a square.

length of the tracer arm times the roll of the measuring wheel. In symbols, $A_r = Ld$, where A_r is the area of the region, L is the length of the tracer arm, and d is the roll of the wheel.

The subscript 'r' is to remind us that A_r is the area of the region being measured (there will be other areas in a moment). Since it will be easier to note the number of rotations made by the measuring wheel, a more useful formula may be $A_r = LnC$, where C is the circumference of the measuring wheel and n is the number of rotations it makes, including fractions.

The proof of the Planimeter Theorem is taken up in the next subsection.

Small Project #3. Use your TinkerToy® planimeter to measure several regions of known area, ranging from 10 to 200 square inches. To do this, you need values for C and L, the circumference of the wheel and the length of the tracer arm, respectively. For C, see the comments in Small Project #1. For L, you can start with 10.5″, the length of the long sticks. If you get results that are larger or smaller than the true area by a consistent percentage, you need to use better values for L and C. The length L of the tracer arm should be the distance on the table between the tracer point and the center of the elbow joint. After calibrating your planimeter, use it to measure some regions of unknown areas. To check your results, compare them to those obtained by others. If you measure the area of a lake, state, or country on a map, convert your result (in square inches) to square miles by multiplying by the *square* of the length scale on the map. For example, if the scale on the map is 15 miles to an inch, you should multiply by $15^2 = 225$, that is, one square inch on the map represents 225 square miles. A lake measuring five square inches on such a map has an area of $5 \times 225 = 1125$ square miles. To check your results, you may be able to find the area in an atlas or almanac.

On a commercial planimeter, the measuring wheel has a scale attached to it, as seen in Figure 9. You do not need to use the formula. Instead, the formula is incorporated into the device, and you read the area directly from the scale. Planimeters are set to measure either in square inches or square centimeters. On some, the length of the tracer arm is adjustable, which allows you to make measurements in either system of units. You can even set it to measure in square miles when using a map with a particular scale.

A linear planimeter is used in exactly the same way. The mechanical difference between the two types is that the "elbow" (the end of the tracer arm opposite the tracer point) of a linear planimeter moves along a straight line, whereas in a polar planimeter it moves along a circle.

Figure 9. Commercial planimeter and the scale on its measuring wheel.

How Planimeters Work: Proof of the Planimeter Theorem

This seems too simple to work, or, if it works, it seems like it must give only an approximation. However, a planimeter gives exact results, if it is exactly manufactured and used correctly. The proof is based on two simple theorems (Theorems 5 and 6 below) about the area swept out by a line segment moving in a plane, which are significant results in their own right.

Theorem 3 (Moving Segment Theorem [2, p. 295; 3, p. 451]**)** *The area A swept by a straight line segment of length L as it moves from P_1Q_1 to P_2Q_2 (Figure 10) is given by $A = Ld$, where d is the component of the distance the midpoint of the segment moves perpendicular to the segment.*

Figure 10. Moving segment sweeping out an area.

It is important to note that d can be (and often is) less than the distance the midpoint moves. To see this, consider the familiar formula for the area of a parallelogram, $A = bh$, where b is the length of the base and h is the height. If a moving segment of length b moves from one base to the other, it sweeps out the parallelogram. The actual distance moved by the midpoint is the length of the other sides of the parallelogram, but the component of its distance perpendicular to the base is the height h. Thus we have $L = b$ and $d = h$, and so the formula for area given by the theorem agrees with that for the area of the parallelogram.

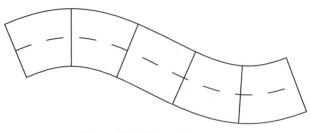

Figure 11. Ribbon Theorem.

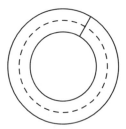

Figure 12. Annulus.

Some readers may be familiar with another special case of Theorem 3, sometimes called the "Ribbon Theorem," illustrated in Figure 11.

Theorem 4 (Ribbon Theorem) *The area of a curved region of constant width equals the width of the region times the length of its median curve (the curve described by the midpoints of its widths).*

A special case of this theorem is the well-known formula for the area of an annulus. Let L be the difference between the radii of the circles (shown as a line segment in Figure 12), let C be the circumference of the median circle (dotted line). Then the area of the annulus is $A = LC$. A bit of elementary algebra and geometry shows that this is equivalent to the "obvious" formula, $A = \pi R^2 - \pi r^2$, where R and r are the larger and smaller radii, respectively.

You have probably already noticed that the formulas in Theorems 2 and 3 are essentially the same, but their interpretations are different. To begin making the connection, note that the d in Theorem 3, awkwardly described as the component of the distance the midpoint moves perpendicular to the segment, is *exactly* the signed distance recorded by a measuring wheel mounted at the midpoint of the tracer arm with its axis parallel to the segment, just as the wheel is attached to the tracer arm of the planimeter. (In the next section we see what happens if the wheel is not mounted at the midpoint.) We note that for Theorem 3 to be true, areas that are swept more than once must count as many times as they are swept. Furthermore, if part of the segment backtracks, that portion of the area must be counted negatively. Like distance, area is not negative, but we can consider a new notion of *signed area* relative to the moving segment. The segment has a forward direction (the direction in which the signed distance measured by the wheel is positive). Signed area swept in that direction is positive. Signed area and signed distance in the opposite direction are negative. Thus Theorem 3 is really about signed area. It can be restated more simply as follows.

Theorem 5 (Moving Segment Theorem, restated) *Suppose a line segment of length L is equipped with a measuring wheel at its midpoint with its axis parallel to the segment. If the segment moves from one position in a plane to another, the signed area swept out by the segment is given by $A = Ld$, where d is the signed distance recorded by the wheel during the motion.*

So now the d in Theorem 2 is the same as the d in Theorem 3. What about the areas? Note that the overall area covered by the tracer arm is typically much larger than the area being measured. However, the *signed* areas are the same. This follows from the following Area Difference Theorem.

Theorem 6 (Area Difference Theorem) *If the endpoints of a moving line segment each move around closed curves, both in the same direction, then the signed area swept by the segment is equal to the difference between the two areas traversed by the endpoints. In symbols, the signed area is $A = A_r - A_\ell$, where A_r and A_ℓ are, respectively, the areas of the regions traversed by the right and left endpoints (Figure 13, lower right).*

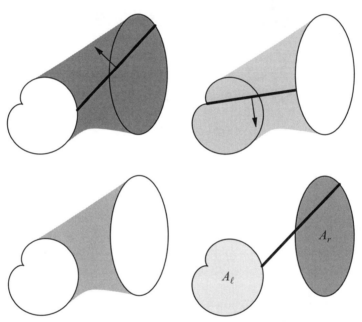

Figure 13. Intuitive proof of the Area Difference Theorem [6].

(Now the subscript 'r' stands for 'right'! Not to worry—we will always take the region measured by a planimeter to be the one on the right.) Intuitively, here is why this is true. The total signed area swept by the segment is equal to the area swept in the positive direction (Figure 13, upper left) minus the area swept in the negative direction (Figure 13, upper right). The positive area has two parts, namely A_r, which is encircled by the right endpoint, and that between the two regions (Figure 13, lower left). Likewise, the negative signed area has two parts, A_ℓ, encircled by the left endpoint, and that between the two regions. When we add to get the total signed area swept by the segment, the parts in common cancel, leaving the signed area swept by the segment as the area encircled by one end minus the area encircled by the other end, that is, $A_r - A_\ell$. For an animation of this, see Robert Foote's website [6]. Bruce Atwood [1] gives a rigorous proof using Green's Theorem.

Combining Theorems 5 and 6, we have $A_r - A_\ell = A = Ld$. If the moving segment is the tracer arm of a planimeter, then A_r is the area enclosed by the tracer point of the planimeter, that is, the area of the region being measured, and A_ℓ is the area enclosed by the "elbow." But, as noted earlier, the linkage of a polar planimeter is designed so that the elbow never leaves the arc of a circle, hence, properly used, it will encircle no area, that is, $A_\ell = 0$. Hence, the area enclosed by the tracer point equals the signed area swept by the tracer arm. The situation for the linear planimeter is similar—the end of the tracer arm that moves along the linear track encloses no area, so again $A_\ell = 0$. In both cases we have that $A_r = Ld$. This completes the proof of Theorem 2, the Planimeter Theorem.

It Does Not Matter Where the Wheel Is

Up to this point it has been assumed that the wheel of the planimeter is located at the midpoint of the tracing arm. You may have noticed, however, that on most commercial planimeters, the wheel is generally *not* at the midpoint. In this section we will see that the wheel can, in fact, be located at any point along the tracing arm.

From the Moving Segment Theorem we have the formula

$$A = Ld, \qquad (1)$$

where A is the signed area swept out by a motion of the tracer arm, d is the corresponding roll of the wheel at the midpoint, and L is the length of the tracer arm. Relocating the wheel changes this formula, but not the formula in the Planimeter Theorem. Again, the formulas are the same, but they have different interpretations. The formula in the Moving Segment Theorem is true regardless of the segment's position, whereas the formula in the Planimeter Theorem is valid only after the region has been measured. We need to determine how locating the wheel at another point changes formula (1).

Small motions of the tracer arm come in two independent types. The general small motion is a combination of these. The first type is translational, when the planimeter's initial and final positions are parallel to each other. In this case the signed area swept out is given by (1), since this is just the formula for the area of a parallelogram, as noted in the previous section. This is valid no matter where the wheel is located along the tracer arm.

The second type of motion is when the tracer arm rotates about the wheel, that is, the wheel does not roll, and the planimeter's initial and final positions make some angle θ to each other (see Figure 14). Two circular sectors are formed. It is important to note that they are swept out in opposite directions, and so one of them contributes positively to the signed area and the other contributes negatively. If the rotation is clockwise, the signed area of the left sector is positive and that of the right sector is negative—the opposite is true for a counterclockwise rotation. Note that if the wheel is mounted at the midpoint of the tracer arm, these two circular sectors are congruent, and so their signed areas cancel, which makes this special case easier.

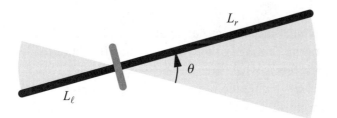

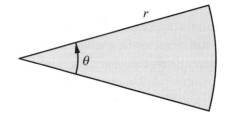

Figure 14. Rotation about the wheel [6].   Figure 15. Circular sector.

Suppose that the wheel is located along the tracer arm so that the length of the arm to the left of the wheel is L_ℓ and the length of the arm to the right of the wheel is L_r. Then L_ℓ and L_r are the radii of the two circular sectors. Recall that a circular sector of radius r and central angle θ (see Figure 15) has area given by

$$A = \frac{1}{2}r^2\theta \quad \text{or} \quad A = \frac{\pi}{360}r^2\theta.$$

Here and in what follows, the first formula is used if the angle θ is measured in radians; the second formula is used if the angle θ is measured in degrees.

Assume for the moment that the small rotation of the tracer arm about the wheel in Figure 14 is counterclockwise. The signed area swept out is the difference between the areas of the circular arcs. Applying the previous formula we have

$$A = \frac{1}{2}L_r^2\theta - \frac{1}{2}L_\ell^2\theta = \frac{1}{2}\left(L_r^2 - L_\ell^2\right)\theta \quad \text{or} \quad A = \frac{\pi}{360}L_r^2\theta - \frac{\pi}{360}L_\ell^2\theta = \frac{\pi}{360}\left(L_r^2 - L_\ell^2\right)\theta. \qquad (2)$$

Note that even though the angle θ is positive, the signed area A can be positive or negative, depending on whether L_r is bigger or smaller than L_ℓ. If we assume the standard convention that counterclockwise angles are positive and clockwise angles are negative, then this formula is also correct for clockwise rotations—if the rotation is changed from counterclockwise to clockwise, both θ and A are multiplied by -1, and so equality is maintained in (2).

We have seen that (1) and (2) give the signed areas swept out by the two types of small motions of the tracer arm. The total signed area swept out is the sum of these:

$$A = Ld + \frac{1}{2}\left(L_r^2 - L_\ell^2\right)\theta \quad \text{or} \quad A = Ld + \frac{\pi}{360}\left(L_r^2 - L_\ell^2\right)\theta. \qquad (3)$$

In fact, this formula is valid for large motions as well.

Theorem 7 (Improved Moving Segment Theorem) *Suppose a line segment of length L is equipped with a measuring wheel with its axis parallel to the segment. If the segment moves from one position in a plane to another, the signed area swept out by the segment is given by formula* (3).

Two observations are in order. First, note that if the wheel is at the midpoint of the tracer arm, then $L_\ell = L_r$ and (3) reduces to the special case (1) of the previous section, as it should. Second, the two terms of the sum in (3) are complementary to each other in a mechanical way: Ld applies when the wheel rolls and the tracer arm does not rotate, whereas $\frac{1}{2}\left(L_r^2 - L_\ell^2\right)\theta$ applies when the tracer arm rotates and the wheel does not roll.

Now suppose the planimeter is used to measure an area. The Area Difference Theorem applies and we have $A = A_r - A_\ell$. As in the previous section, A_ℓ is zero because the left endpoint of the arm does not go around any area, and so A equals A_r, the area of the region being measured. In addition, note that the tracer arm does not make a full rotation; its direction simply oscillates around its initial direction. It follows that the net change in angle, θ, is zero. Plugging these into (3) we get

$$A_r = Ld.$$

This is the same result as in the Planimeter Theorem. Since it does not involve L_ℓ and L_r, it follows that the wheel can be located anywhere along the tracer arm. The area of the region is the product of the length of the tracer arm and the amount the wheel rolls, and this is what is read on the wheel's scale.

Commercial planimeters are invariably used by tracing the region clockwise, as opposed to counterclockwise. This makes the wheel roll in the opposite direction. If you have access to one, you will note that the scale on the wheel is calibrated so that it *decreases* in what we have used as the forward direction.

The Prytz Hatchet planimeter

We are now in a position to understand the operation of a very simple and curious type of planimeter, known as a hatchet planimeter. It has no moving parts! It is surprising that it measures anything at all, but with formula (3) in the Improved Moving Segment Theorem we will be able to show that it measures area, at least approximately, and that the error has a very nice geometric interpretation.

Area Without Integration: Make Your Own Planimeter

Figure 16. Hatchet planimeter [11].

This type of planimeter was invented in about 1875 by Holger Prytz, a Danish cavalry officer and mathematician, as an economical and simple alternative to Amsler's polar planimeter. It consists of a rod with its ends bent at right angles to its length. One end, labeled T in Figure 16, is sharpened to a point. The other end, labeled C, is sharpened to a chisel edge parallel to the length of the rod. The chisel edge is slightly rounded, making it look somewhat like a hatchet (hence the name). Prytz referred to it as a "stang planimeter," "stang" being the Danish word for "rod."

The first author had a Prytz planimeter made from a used chemistry ring stand in the Wabash College maintenance shop. It is also easy to make one out of TinkerToys® or wooden dowels, as in Figure 17. Use two 1/4" dowels for the vertical pieces. Sharpen one to a point in a pencil sharpener. The chisel edge can be fashioned in the other dowel by carving it. Alternatively, cut a thin slot in the end and wedge into it a small piece of sheet metal or the tip of an old table knife. Even a sturdy piece of thin cardboard will work (cut a small circle from an old playing card). We have even seen one in which the chisel edge is a circular, rotating, pizza slicer with its handle extended by a dowel. The chisel edge does not need to be terribly sharp, but it does need to be rounded and stiff enough that it will track correctly. It is important to be able to hold and guide the device with a relaxed grip. It helps if the center of gravity is low. The vertical pieces should be no longer than necessary and the horizontal bar should be lightweight, say a 3/8" dowel or other piece of wood. A small weight (a coin, for instance) can be added just above the chisel edge to help it track better (see Figures 17 and 19).

Figure 16 illustrates how the device is used. The end sharpened to a point is the tracer point T. Before beginning to trace the boundary of the region to be measured, note the initial location C_i of the chisel edge.

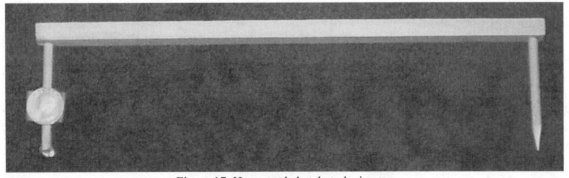

Figure 17. Homemade hatchet planimeter.

Then trace the boundary of the region, taking care not to apply any torque to the rod. Using a relaxed grip is important here. The chisel edge will track better if it drags on a piece of paper as opposed to a smooth counter top. The chisel edge will describe a zig-zag path that is always tangent to the direction of the rod (see Figures 16 and 18). After traversing the boundary of the region, make a note of the final location C_f of the chisel edge. Measure the distance between C_i and C_f and call it D. Measure the length L of the rod, that is, the distance between the tracer point and the point where the chisel edge contacts the paper. Then the area A_r of the region is given, at least approximately, by their product,

$$A_r \approx LD. \tag{4}$$

What could be simpler?

This may seem too good to be true, and in some sense it is. Of course, *every* measurement is an approximation, but the main source of error in the use of a standard polar planimeter is due to the errors made by the user (not following the curve accurately or not reading the scale correctly). On the other hand, the error in the approximation made by a hatchet planimeter is inherently part of the device. The error can be significant, even if the user has a steady hand and measures D and L accurately. With the background of the previous sections, we can understand this approximation and the nature of the error, which can lead to a more accurate use of the instrument.

> *Small Project #4.* Make a hatchet planimeter. Use it and formula (4) to measure the areas of several regions, and compare your results with those you obtained with the polar planimeter. You may want to repeat this project after reading the rest of this section to understand how to minimize the error inherent in the hatchet planimeter.

The hatchet planimeter rod is the analog of the tracer arm in a polar or linear planimeter; it is the moving line segment that sweeps out signed area. As in the previous sections, take the tracer point to be the right end point. Then the chisel edge is the left end point. We would like to apply the Improved Moving Segment Theorem and the Area Difference Theorem, however there are two obstacles, the most obvious of which is that there is no wheel. The second is that the chisel edge does not traverse a closed curve, which is required by the Area Difference Theorem.

We overcome these obstacles with a "thought experiment." Consider using the hatchet planimeter to measure the region shown in Figure 18. The initial and final positions of the rod are shown, as well as the path of the chisel. After the tracer point returns to its starting position, imagine holding the tracer point fixed and rotating the rod from its final position back to its initial position. The chisel edge follows a circular arc from C_f to C_i. The radius of the arc is L, the length of the rod, and its center is the base point B where the tracing starts and stops. Now the chisel edge has gone around a closed curve, and the Area Difference Theorem applies. Actually we need a modification of the Area Difference Theorem because the curve intersects itself.

Theorem 8 (Improved Area Difference Theorem) If the endpoints of a moving line segment each move around closed curves, then the signed area swept by the segment is $A = A_r - A_\ell$, where A_r and A_ℓ are, respectively, the signed areas of the regions traversed by the right and left endpoints.

It is the same as before, but with a more sophisticated interpretation. The area of a region is taken to be signed (positive or negative) depending on the direction its boundary is traversed. The area is taken to be positive when the boundary is traversed counterclockwise; it is taken to be negative when the boundary is traversed clockwise.

The tracer point has gone around its region counterclockwise, and so A_r is simply the area we are measuring. The chisel edge, however, has gone around two regions that are roughly triangular (shaded in Figure

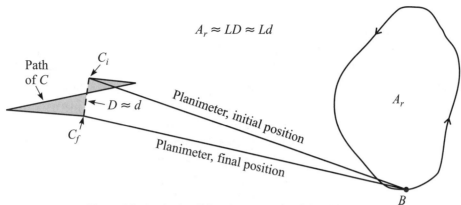

Figure 18. Analysis of the zig-zag path of the chisel [6].

18). In particular it has gone around the upper one (with C_i as a vertex) clockwise and the lower one (with C_f as a vertex) counterclockwise. Let A_{cw} and A_{ccw} be, respectively, the areas of the upper and lower triangular regions. Then A_ℓ, the signed area enclosed by the path of the chisel, is $A_\ell = A_{ccw} - A_{cw}$. By the Improved Area Difference Theorem, the signed area swept out by the rod is

$$A = A_r - A_\ell = A_r - (A_{ccw} - A_{cw}). \tag{5}$$

In a more complicated example there could be many triangular regions, in which case A_{cw} would be the total area of those traversed clockwise, and similarly A_{ccw} would be the total area of the those traversed counterclockwise.

Continuing our thought experiment, imagine that the chisel is replaced by a wheel with its axis parallel to the rod as usual. How much does the wheel roll? While the tracer point goes around the boundary of the region being measured and the wheel follows the zig-zag path, *the wheel does not roll at all*. This is because the direction of motion is parallel to the rod, and this is the direction in which the wheel only slides. On the other hand, when the planimeter rod is rotated from its final position back to its initial position, the amount it rolls is the length of the circular arc from C_f to C_i. As in the previous sections, denote the roll of the wheel by d.

We can now apply formula (3) from the Improved Moving Segment Theorem. Just as in the previous section, since the planimeter (in the thought experiment) returns to its initial position and does not make a full rotation, the net change in angle, θ, is zero. Plugging this and (5) into (3), we have

$$A_r - (A_{ccw} - A_{cw}) = Ld.$$

Solving for the area we are interested in, we get

$$A_r = Ld + (A_{ccw} - A_{cw}).$$

If we think of the product Ld as an approximation for A_r, then this formula shows that the error made by the approximation is the quantity in parentheses, that is,

$$A_r \approx Ld \quad \text{and} \quad \text{error} = A_{ccw} - A_{cw}. \tag{6}$$

Note that d is approximately the distance D between C_f to C_i that we measured earlier in (4). The use of D instead of d (which is how (4) follows from (6)) changes the error, but not significantly. However this had a significant effect on the history of the instrument, as we will see in a moment.

The analysis leading to (6) shows that the amount of error made is a trade-off between the triangular regions traversed clockwise by the chisel and those traversed counterclockwise. With some experimentation you can convince yourself that the starting position of the planimeter affects this trade-off. With practice you can judge which positions are better than others. In a more detailed analysis, Prytz showed that the error is substantially reduced, although not entirely eliminated, by starting the tracing point at the centroid, or center of mass, of the region, as is suggested in Figure 16 (the centroid is labeled B_c). Draw a line segment from the centroid to some point on the boundary. Start tracing at the centroid, following the line segment out to the boundary. After going around the boundary, follow the line segment back to the centroid. Of course determining the exact location of the centroid is harder than finding the area, but the idea is that an educated guess should reduce the error. Prytz also showed that the error is smaller if a longer planimeter is used. Prytz' paper is quite technical, but its main points are summarized in [5].

The hatchet planimeter has an amusing history. For some unknown reason, Prytz wanted to remain anonymous, at least initially. He published the theory of how the instrument works (considerably more complicated than that presented here) under the pseudonym "Z" in 1886. Eventually others made design modifications and wanted to take credit for the device. In 1894–96 and again in 1906–07 there ensued a lively debate in the pages of the journal *Engineering* as to the identity of the inventor and the merits of the subsequent modifications [5, 10].

Most of the design changes involved measuring the circular arc distance d between the initial and final chisel locations instead of simply using the straight line distance D. The inventors were under the mistaken impression that this would correct the inherent error of the device. (It does not, as the analysis above shows.) One such modification, by an engineer named Goodman in 1896, is shown in Figure 19. Here a curved scale is incorporated into the planimeter rod. The radius of the curve is the same as the length L of the rod. By placing the curved scale alongside the initial and final chisel locations, one measures d directly. As with the scale on the wheel of a polar planimeter, Goodman incorporated the constant multiplier L so that the product Ld, the approximate area, is read from the scale. Note the weight over the chisel edge to help it track better. The design changes made the instrument more expensive and defeated its purpose, in Prytz's opinion, as an economical alternative to a polar planimeter. In harsh criticism of the other inventors' designs, Prytz advised engineers [12] "rather than use the 'improved stang planimeters,' let a country blacksmith make them a copy of the original instrument."

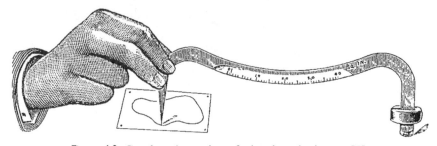

Figure 19. Goodman's version of a hatchet planimeter [7].

Conclusion

This simple analysis of how these planimeters work is due to Henrici [8], who gives a very complete and interesting history of planimeters up through 1893. For more details and references on the history of hatchet planimeters, see [10] and [5]. Another good reference on planimeters and other mathematical devices, which may be easier to find, is the book by Murray [9]. A recent article [14] gives very nice proofs of how polar and linear planimeters work, suitable for students who have studied calculus.

Polar planimeters are readily available on eBay [4], and can typically be purchased for $50 to $100. Linear planimeters appear occasionally and are generally more expensive. Manufactured hatchet planimeters are quite rare and expensive. You might also find a planimeter hiding forgotten in a drawer or closet in a local school or college science department, or in the attic of a retired scientist, architect, or engineer. There are other kinds of planimeters. For further information and pictures, see the first author's web page [6].

Planimeters are still manufactured, and a web search will turn up a few companies that sell them. Many have electronics added, but mechanically they are really just polar and linear planimeters with calculators riding piggyback; Amsler would easily recognize them. The only difference is that the roll of the wheel is read electronically instead of from a scale, and the calculator allows the user to convert between different units, say from square inches to square centimeters, or to square miles based on the scale of a map. Engineering, architecture, and natural resource students are still taught how to use them at many schools, and they are used by many professionals in these and other fields.

Areas are useful in numerous applications. Uses of planimeters range from finding areas of lakes on maps, to finding areas of tumors on x-rays and of leaves and flower petals. The Science Museum in London has a huge polar planimeter with arms that are over three feet long that was used in the leather industry to measure areas of hides.

Small Project #5. Now that you know how planimeters work, here are some related things to explore.

- Make a more accurate polar planimeter than the one you made out of TinkerToys®.
- Make a linear planimeter.
- Determine how the accuracy of a Prytz planimeter depends on its starting orientation. For a given starting point along the boundary of the region, some orientations will minimize the error made in (6).
- Do your science departments have a planimeter tucked away in a drawer? Does your local historical society have a planimeter in its collection? Do they have it on display? Do they know what it is and how to use it?
- Do a web search to find additional uses of planimeters. You might even find the web page in which some researcher in food science has used a planimeter to measure the area of a cookie!

Acknowledgments

This paper began as a handout for a presentation made by the second author in January 2004 at an MAA Short Course on the history of mathematical instruments. We would like to express thanks to the editor, Amy Shell-Gellash, who was also one of the organizers of the course, for encouraging us to expand it into a paper.

Figures 3 and 4 are from *Mathematical Machines*, Vol. 2, *Analog Devices* by Francis J. Murray [9], Copyright ©1961 by Columbia University Press. Reprinted with permission of the publisher.

References

1. B. Atwood, *Area Measurement: Planimeters & Green's Theorem*, http://www.attewode.com/Calculus/AreaMeasurement/area.htm, 28 August 2007.
2. R. Courant, *Differential and Integral Calculus*, II. Originally published in 1934 as *Vorlesungen über Differential- und Integralrechnung*. Translated by E. J. McShane. Wiley Interscience, New York, 1988.
3. R. Courant and F. John, *Introduction to Calculus and Analysis*, II, Wiley Interscience, New York, 1974.
4. eBay on line auctions, www.ebay.com.
5. R. Foote, "Geometry of the Prytz Planimeter", *Reports on Mathematical Physics*, **42** (1998) 249–271.
6. ——, *Planimeters*, http://persweb.wabash.edu/facstaff/footer/planimeter/planimeter.htm, 28 August 2007.
7. J. Goodman (pub. anon.), "Goodman's Hatchet Planimeter", *Engineering*, Aug. 21, 1896, 255–56.

8. O. Henrici, *Report on Planimeters*, British Assoc. for the Advancement of Science, Report of the 64th meeting, (1894) 496–523.

9. F. J. Murray, *Mathematical Machines*, Vol. 2, *Analog Devices*, Columbia University Press, New York, 1961.

10. Olaf Pedersen, "The Prytz Planimeter", in *From Ancient Omens to Statistical Mechanics*, J. L. Berggren and B. R. Goldstein, eds., University Library, Copenhagen, 1987.

11. A. Poulain, "Les Aires des Tractrices et le Stang-Planimètre", *J. de Mathématiques Spéciales*, Vol. 4, No. 2 (1895) 49–54.

12. H. Prytz, "The Prytz Planimeter", (two letters to the editor), *Engineering*, September 11, 1896, 347.

13. Würzburger Mathematikausstellungen: Kegel-Planimeter (Würzburger Mathematics exhibition: Cone Planimeter), http://www.didaktik.mathematik.uni-wuerzburg.de/History/ausstell/planimet/gonnella.html, 28 August 2007.

Added in proofs:

14. T. Liese, "As the Planimeter's Wheel Turns: Planimeter Proofs for Calculus Class", *The College Mathematics Journal*, 38 (1007) 24–31.

Historical Mechanisms for Drawing Curves

Daina Taimina
Cornell University

Introduction

If you have a collection of straight sticks that are pinned (hinged) to one another, then you can say you have a linkage like in the windshield wipers in your car or in some desk lamps. Linkages can also be robot arms. It is possible that our own arms caused people to start to think about the use of linkages.

In this paper I will discuss how linkages and other historical mechanisms (that involve sliding in groove or rolling circles) can be used for drawing different curves and in engineering to design machine motion. This knowledge was very popular at the end of the 19th century, but much of it was forgotten during most of the 20th Century. Now there is, among mathematicians and engineers, renewed interest in these mechanisms and in kinematics — the geometry of pure motion. Study of these mechanisms can be used in classrooms as a way to show interconnections between mathematics and technology and provide a bridge to interesting history that can bring meaning into the classroom. For examples, Descartes considered only those curves that could be drawn with mechanical devices. Curves were constructed from geometrical actions, many of which were pictured as mechanical apparatuses. After curves had been drawn, Descartes introduced coordinates and then analyzed the curve-drawing actions in order to arrive at an equation that represented the curve. Equations did not create curves; curves gave rise to equations. [10]

Some early mechanisms

We can find linkages in drawing curves in ancient Greece. Mechanical devices in ancient Greece for constructing different curves were invented mainly to solve three famous problems: doubling the cube, squaring the circle and trisecting the angle. There can be found references that Meneachmus (~380–~320 B.C.) had a mechanical device to construct conics which he used to solve the problem of doubling the cube. One method to solve problems of trisecting an angle and squaring the circle was to use the quadratrix of Hippias (~460–~400 B.C) — this is the first example of a curve that is defined by means of motion and can not be constructed using only a straightedge and a compass. Proclus (418–485) also mentions Isidorus from Miletus who had an instrument for drawing a parabola [13, p. 58]. We can not say that those mechanical devices consisted purely of linkages, but it is important to understand that Greek geometers were looking for and finding solutions to geometrical problems by mechanical means. These solutions mostly were needed for practical purposes, for example, building different structures. This was in spite of the fact that in theoretical Euclidean geometry there existed demands for pure geometrical constructions using only "divine instruments" — compass and straightedge.

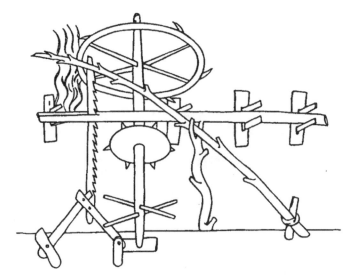

Figure 1. 13th century sawmill, sketched by Villard de Honnecourt
(note the linkage in the lower left-hand corner). [24]

We can find use of linkages in some old drawings of machines like in Figure 1. But in the 16th century the largest account on Renaissance engineering was done by Agostino Ramelli (1531–c.1610) who in 1588 in Paris published the volume "The diverse and artifactitious machines of captain Agostino Ramelli".

Another source for the use of machines in the 16th century are the works of Georg Bauer, better known by the Latin version of his name Georgius Agricola (German, 1494–1555), who is considered a founder of geology as a discipline. Agricola's geological writings reflect an immense amount of study and first-hand observation, not just of rocks and minerals, but of every aspect of mining technology and practice of the time.

> I have omitted all those things, which I have not myself seen, or have not read or heard of from persons upon whom I can rely. That which I have neither seen, nor carefully considered after reading or hearing of, I have not written about. The same rule must be understood with regard to all my instruction, whether I enjoin things, which ought to be done, or describe things, which are usual, or condemn things, which are done. [1]

In the drawings that accompanied Agricola's work we can see linkages that were widely used for converting the continuous rotation of a water wheel into a reciprocating motion applied to piston pumps. [1]

In the past linkages could be of magnificent proportions. Linkages were used not only to transform motion but also were used for the transmission of power. Gigantic linkages, principally for mine pumping operations, connected water wheels at the riverbank to pumps high up on the hillside. One such installation (1713) in Germany was 3 km long. In Figure 2 we can see the use of a linkage in a 16th century sawmill (See the six connected diagonal bars at the top of the frame — this linkage is today often called "lazy tongs".) Figure 2 is contained in a 16th century book Theatrum instrumentorum et Machinarum [5] (available on-line), which describes many earlier mechanisms.

Early curve drawing

In the Renaissance, ancient methods of constructing different curves were not satisfactory. Renaissance engineers needed new mechanisms that would trace precise trajectories, so that they could be used, for example, to drive the cutters for making precision lenses, gears, and guides for mechanical motion. For example, Leonardo da Vinci (1452–1519) invented a mechanism that could be adapted on a lathe for cutting

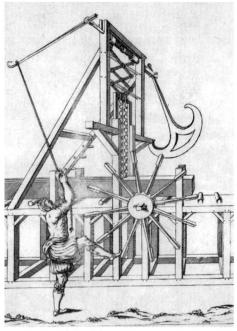

Figure 2. Besson's drawing of a sawmill [5]

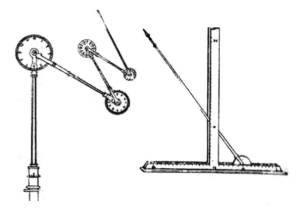

Figure 3. Dűrer's curve drawing devices [12]

parts with elliptic cross-sections. See [29], model D-10. He had ideas about several other mechanisms that would trace various mathematical curves. Mechanical devices for drawing curves were used also by Albrecht Dűrer (1471–1528). See Figure 3 for two pictures from Dűrer's "Four books on proportions" in which he describes their uses.

When Rene Descartes (1596–1650) published his *Geometry* (1637) he did not create a curve by plotting points from an equation. He always first gave geometrical methods for drawing each curve with some apparatus, and often these apparatus were linkages. See Figure 4. This tradition of seeing curves as the result of geometrical actions can be found also in works of Roberval (1602–1675), Pascal (1623–1662) and Leibniz (164–1716). Mechanical devices for drawing curves played a fundamental role in creating new symbolic languages (for example, calculus) and establishing their viability. The tangents, areas and arc lengths associated with many curves were known before any algebraic equations were written. [10]

In this paper we are mostly describing mechanisms that were of interest to engineers and are connected with the Kinematic Models for Design Digital Library (KMODDL) collection [29]. For an excellent collection of mechanisms related to the history of mathematics, see [39], which discusses curve-drawing and other mathematical devices together with their history, models, animations, and bibliography.

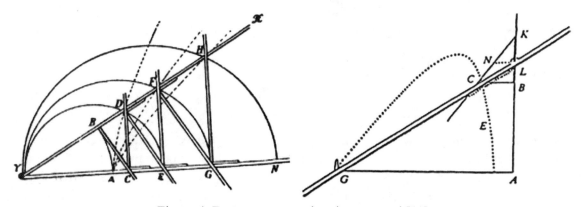

Figure 4. Descartes curve drawing apparati [11]

Drawing ellipses

The most familiar mechanical construction of an ellipse is the "string" construction, which dates back to ancient Greece, if not even earlier. It is based on the two-focus definition of an ellipse. As Coolidge (1873–1958) in [7] states:

> The Greeks must have perceived that if the two ends of a piece of string be made fast, at two points whose distance apart is less than the length of the string, the locus of a point which pulls the string taut is an ellipse, whose foci are the given points.

A further modification given by Coolidge is the following:

> We have three pins placed at the centre and foci of the curve. The ends are knotted together and held in one hand. From here the string passes around the centre pin on the left and right under two foci, and is drawn tight by a pencil point above. If the hand holding the two ends remain fixed, the pencil point will move along an arc of an ellipse. By changing the position of the fixed hand this may be altered at pleasure, except that the foci are fixed. But if a small loop be made around the pencil point so that it cannot slip along the string, and the hand holding the ends be pulled down, then the pencil point will trace an arc of hyperbola, for the difference in the lengths of the two path from the hand to the point is constant, and so is the difference of the distances to the two foci. (See Figure 5.)

The first who wrote about this construction of an ellipse by means of a string was Arab mathematician Abud ben Muhamad. His works date to the middle of the 9th century. See [7].

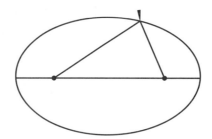

Figure 5. Using a string to draw an ellipse

Another device for constructing an ellipse can be based on a fact that if all chords of a circle having a certain direction are shrunk or stretched in a constant ratio, the resulting curve is an ellipse. A description of such an "ellipsograph" can be found in [13, p. 228].

Rolling circles produce straight lines and ellipses

In mechanics when obtaining an elliptic path for machine motion, La Hire's (1614–1718) theorem is often used:

La Hire's Theorem. *If a circle rolls, without slipping, so that it is constantly tangent internally to a fixed circle of twice its radius, the locus of a point on its circumference is a diameter of the fixed circle, while the locus of a point rigidly attached to it elsewhere is an ellipse.* [23, p. 351] (See Figure 6.)

La Hire's proof of this theorem was based on the theory of rolling curves or centrodes. But this result was proved before La Hire. See [35, p. 130] for a description and indication of proof by Proclus of this theorem. Note that this mechanism is not a linkage.

There is another theorem that has been rediscovered many times, but the priority seems to go to Nasir al-Din al-Tusi (1201–1274). Al-Tusi made the most significant refinements of Ptolemy's model of the plan-

Figure 6. Tusi-couples [29, Reuleaux models S-02 and S-04] (photo Prof. F. C. Moon)

etary system up to the development of the heliocentric model by Copernicus (1473–1543). Copernicus referred to a famous "Tusi-couple", which resolves linear motion into the sum of two circular motions. The Tusi couple is obtained by rolling a small circle inside a circle of twice the radius. See Figures 6 and 7. The result is a line segment equal to the radius of the largest circle. Mathematicians sometimes call this a *2-cusped hypocycloid*, but engineers refer to such motion as special cases of planetary gears. See [8] and [31]. F. Reuleaux (1829–1905) also used this idea in constructing one of his mechanisms, see Figure 6.

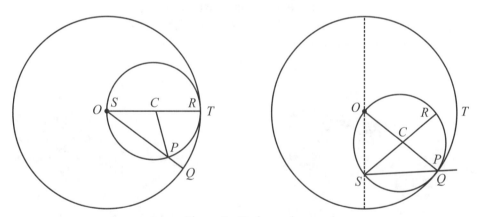

Figure 7. Tusi-couple

In Figure 7, O is the center of the fixed outer circle and C is the center of the rolling inner circle. We start, in the left-hand figure, with the diameter SR of the inner circle coinciding with the radius OT of the outer circle. By hypothesis, $OT = 2CR$. The central angle $\angle PCR$ is twice the subtending angle $\angle PSR$. Thus arc PR = arc QT, since the outer circle is twice the inner circle. Thus, when the inner circle rolls inside the outer circle, the point P will land on the point Q, as in the right hand figure where clearly the arc OS = arc PR. Since arc PR = arc QT, then R must lie on the line OT. Since $\angle SOR$ subtends a diameter, OS is perpendicular to OT. Thus S traces the vertical diameter and R traces the horizontal diameter of the outer circle. This last property leads to the trammel, which is described in the next section.

Trammel for drawing ellipses

The most systematic and complete discussion of the treatment of the conics is found in the *Elementa Curvarum Linearum*, of Johan de Witt, which appeared as an appendix to van Schooten's second Latin edition of Descartes' *Geometrie*, 1659–1661. [15]. Johan de Witt (1625–1672) was a Dutch statesman with considerable skill as a mathematician. While studying law at the University of Leiden he became friends with Francis van Schooten the younger (1615–1660) and received from him an excellent training in Cartesian mathematics. Van Schooten was the main popularizer of Descartes' *Geometrie* in Europe. According to van Schooten, de Witt's treatise was written some ten years prior to the *Geometrie*'s publication.

De Witt describes two more constructions of an ellipse. One of them is the trammel construction, which was described by Proclus, but is also attributed to Archimedes (287–212 B.C.). The trammel is the simplest mechanism for drawing ellipses. It consists of two parts: the fixed frame and the moving coupler rod. See Figure 8. Again this mechanism is not a linkage.

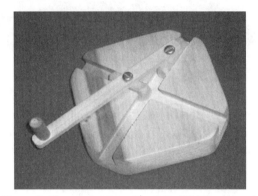

Figure 8. Trammel (photo Prof. D. W. Henderson)

This construction with pictures from Reuleaux kinematic model collection is described in F. Moon's tutorial "How to draw an ellipse" [34]. See Figure 9. In Figure 9 (based on a mechanism used in the Renaissance for cutting precise elliptic lenses) parts of the trammel can be seen in the left-hand picture.

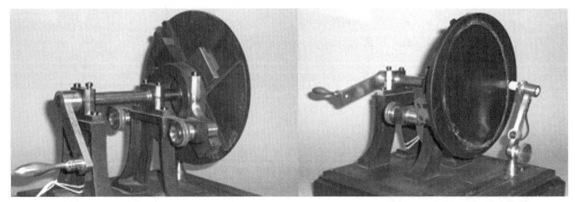

Figure 9. Trammel from Reuleaux's kinematic model collection (photo Prof. D. W. Henderson)

Here is De Witt's proof of why the trammel describes an ellipse.

Given two perpendicular lines AA' and BB' intersecting at O. In a trammel, segment CD moves in a way that C is always on AA' but D is always on BB'. Then if a fixed-point P is chosen on CD (or an extension of this segment), point P describes an ellipse with axis AA' and BB'. When C is at O, then P is in B, and when D is in O, P is in A, thus defining semi-minor and semi-major axis of

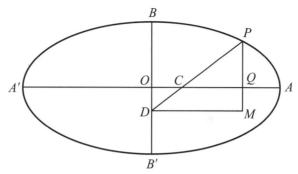

Figure 10. DeWitt's proof of the trammel

the ellipse. Let us draw $PM \perp OA$ and $DM \perp PM$. Then $PQ/PM = PC/PD$ (from similar triangles). But $PC = OB = OB'$ and $PD = OA = OA'$, so $PQ^2/PM^2 = OB^2/OA^2$. But $PM^2 = OA^2 - OQ^2 = (OA - OQ)(OA + OQ) = AQ \cdot A'Q$, and from here $PQ^2 = (OB^2/OA^2)(AQ \cdot A'Q)$, which is the equation of an ellipse if in modern notation we denote AA' and BB' as x- and y-axes: $y^2 = b^2(a^2 - x^2)/a^2$ or $x^2/a^2 + y^2/b^2 = 1$. Leonardo da Vinci suggested using this trammel construction also in a case when AA' and BB' are not perpendicular.

Drawing hyperbolas and parabolas with links and sliders

De Witt also suggested a mechanical construction of a hyperbola in terms of a rotating line and a sliding segment.

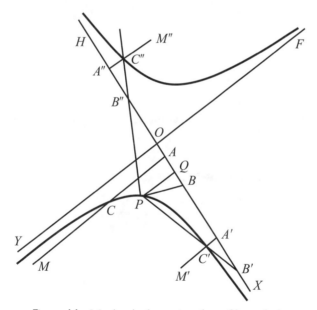

Figure 11. Mechanical construction of hyperbola

Let a line rotate about a fixed point P within the two fixed lines HOX and FOY, intersecting HOX at B. See Figure 11. When the rotating line is parallel to FOY, it intersects HOX at Q and, for each point B, A is chosen on HOX so that $AB = OQ$. At A, draw a line AM parallel to FOY. Let C be the intersection of AM with PB. We want to prove that the locus of the point C is a hyperbola. Triangles PQB and CAB are similar, thus $CA/AB = PQ/QB = PQ/OA$. Since $OQ = AB$, $CA \cdot OA = PQ \cdot OQ =$ a constant; and thus the locus is a hyperbola. [14]

In Figures 12, 13, 14 we can see some well-known linkages with sliders inside for mechanically constructing conics. [36]

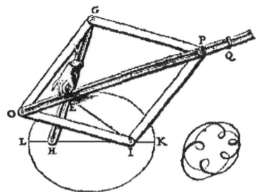

Figure 12. van Schooten mechanism for constructing ellipse

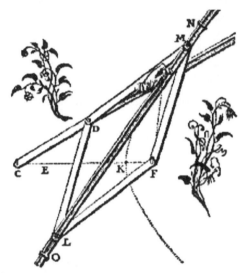

Figure 13. van Schooten mechanism for constructing hyperbola

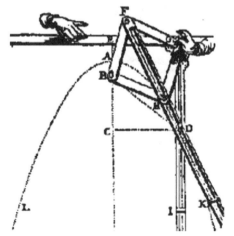

Figure 14. van Schooten mechanism for constructing parabola

To see these mechanisms in action and why they work, see [9], [10]. David Dennis has written about these devices and about their uses in the classroom in [10] and other papers, so I am omitting discussion about them here.

Linkages that draw conic sections

The Hart's Crossed Parallelogram in Figure 15 is a four-bar linkage (with one bar AB attached to the plane) such that $AC = BD$ and $AB = CD$, thus it forms an isosceles trapezoid. The symmetry is preserved as the linkage moves.

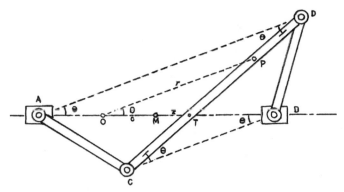

Figure 15. Hart's Crossed Parallelogram [41]

We select a given point P on the traversing bar and draw the line OP parallel to AD and BC. It is clear that OP remains parallel to these lines and O is thus a fixed point of the line AB, because $AO = DP$. Let $AC = BD = 2b$; $AB = CD = 2a$, where $a > b$. Let $OP = r$; $OM = c$, $MT = z$, where M is the midpoint of AB and $\angle POB = \theta$. Then from the Figure 15 (using the symmetry):

$$r = 2(c+z)\cos\theta;$$
$$BC = 2\,(BT)\cos\theta = 2(a-z)\cos\theta;$$
$$AD = 2(AT)\cos\theta = 2(a+z)\cos\theta;$$
$$(BC)(AD) = 4(a^2 - z^2)\cos^2\theta = 4(a^2 - b^2).$$

Combining this result with the first equation to eliminate z, we have

$$a^2\cos^2\theta - (r/2 - c\cos\theta)^2 = a^2 - b^2.$$

This is the polar equation (with origin O) of the path of P. The quantity c is determined, of course, as soon as the point P is selected.

If we invert this curve, taking O as the center of inversion (so that the transformation is $rs = 2k^2$), we obtain

$$a^2 s^2 \cos^2\theta - (k^2 - cs\cos\theta)^2 = s^2(a^2 - b^2).$$

This inverted curve is a conic section, which may more easily be recognized by transferring to rectangular coordinates, using

$$s\cos\theta = x; \quad s\sin\theta = y; \quad s = x^2 + y^2.$$

Thus, we have

$$(c^2 - b^2)x^2 + (a^2 - b^2)y^2 - 2ck^2 x + k^4 = 0.$$

Now, since $a > b$, the coefficient of y^2 is positive and the character of the conic is determined entirely by the coefficient of x^2. Thus the curve is a parabola if $c = b$; an ellipse if $c > b$; and a hyperbola if $c > b$. In Figures 16–18 we have arbitrarily taken $a = 2b$. Thus, the point P' (the inverse of P under the inversion) traces the conic.

Figure 16 shows the linkage for a parabola with $a = 2b = 2c$. Thus, $PD = AO = b$. The point P is inverted to P' by means of Peaucellier cell where $(OE)^2 - (PE)^2 = 2k^2$. For a description of the Peaucellier cell

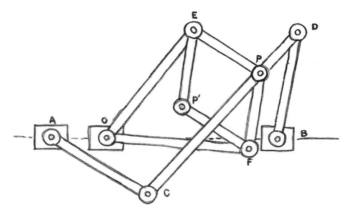

Figure 16. Linkage that draws a parabola [41]

and its properties, see [21, Chapter 16] and [29, tutorial: "Mathematical Tutorial of the Peaucellier-Lipkin Linkage"].

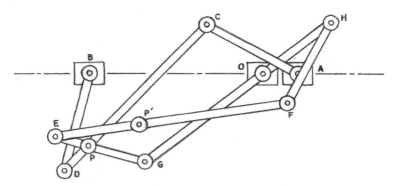

Figure 17. Linkage that draws an ellipse [41]

In Figure 17, the linkage is arranged for an ellipse, where $2a = 4b = 3c$. For the sake of variety, P is inverted to P' by the Hart cell $EFGH$. For a description of the Hart cell and its inversive properties, see [28].

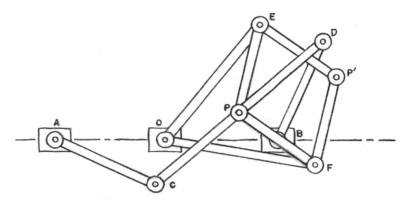

Figure 18. Linkage that draws a hyperbola [41]

Figure 18 gives the arrangement for a hyperbola where $a = 2b$, $c = 0$. (P is midpoint of CD.) These linkages can be built using rigid plastic rods and appropriate connectors (like eyelets), but they can also be investigated with Geometer's Sketchpad. Such investigations can give rise to independent research projects in which the students determine the singular points in these mechanisms and why they occur.

Drawing higher order curves

Examining the theory of algebraic curves of the third degree, Isaac Newton (1643–1727) proposed a mechanism for the generation of circular unicursive curves of third degree, using a four-link chain with two sliding pairs. In Figure 19, see realization of this idea in 20th century as Boguslavskii's conicograph; see [3, p.70] for further descriptions and proofs.

Alfred Bray Kempe (1849–1922) also formulated a famous theorem, which asserts that any algebraic curve can be generated by an appropriate linkage. Some people simplify this by saying that it is possible to design a linkage that will sign your name (as long as your signature is the union of algebraic curves). A proof by Artobolevskii of Kempe's theorem can be found in [3, p.8–12]: He reduces the desired linkage to a series of mathematical operations fulfilled by the individual linkages, which are joined together into the general chain of linkages. Those individual linkages are:

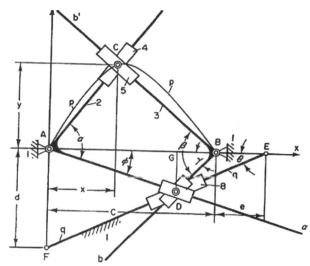

Figure 19. Boguslavskii's conicograph [3]

1. a linkage for conveying a point along the given straight line;
2. a linkage for projecting a given point on to a given line;
3. a linkage which cuts off equal segments on two axes;
4. a linkage for causing a straight line to pass through a given point and be parallel to a given line;
5. a linkage for obtaining proportional segments in two straight lines passing through a given point (multiplying linkage);
6. a linkage for addition of two given segments (summing linkage).

See [26, p. 260–261] for another proof of Kempe's Theorem.

In 1877, A. B. Kempe published a small book: *How to Draw a Straight Line: A Lecture on Linkages* ([28], available on-line at [29]). In this book he describes his theorem and also mentions the work of J. Watt (1736–1819), J. J. Sylvester (1814–1897), Richard Roberts (1789–1864), P. L. Chebyshev (1821–1894), Harry Hart (1848–1920), William Kingdon Clifford (1845–1879), Jules Antoine Lissajous (1822–1880), Samuel Roberts (1827-1913), and Arthur Cayley (1821–1895). This leads to the next section.

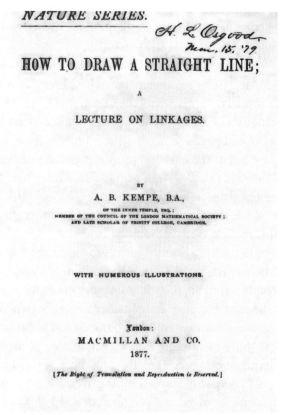

Figure 20. Title page of Kempe's book

Other directions in the theory of linkages

Linkages are closely related with kinematics or the geometry of motion. At first, it was the random growth of machines and mechanisms that put the pressure of necessity to the study of kinematics. Much later, algebraic speculations on the generation of curves were applied to physical problems. Two great figures appeared in the 18th century, Leonard Euler (1707–1783) and James Watt (1736–1819). Although their lives overlap, there was no known contact between them. But both of them were involved with "geometry of motion". Euler's "Mechanica sive motus scientia analytice exposita" (1736–1742) is, to quote Lagrange, "the first great work in which analysis is applied to the science of movement." The fundamental idea of kinematic analysis stems from Euler. Watt, instrument maker and engineer, was concerned with the synthesis of movement. Mechanism designers before Watt had confined their attention to the motions of links attached to the frame. It was Watt who, in 1784, focused on the motion of a point in the middle of a link in a four-bar mechanism. The application of this novel thought allowed Watt to build a double-acting steam engine; which employed a linkage able to transmit force in two directions instead of only one. Euler's theoretical results went unnoticed by engineers for another century, but engineers and mathematicians were devising linkages to compete with or supersede Watt's mechanism. For more information see the online [17].

In 1775 Euler wrote:

> The investigation of the motion of a rigid body may be conveniently separated into two parts, the one geometrical, the other mechanical. In the first part, the transference of the body from a given position to any other position must be investigated without respect to the causes of motion, and must be represented by analytical formulae, which will define the position of each point of the body. This investigation will therefore be referable solely to geometry, or rather to stereotomy. [16, page viii]

It is clear that by the separation of this part of the question from the other that belongs properly to mechanics, the determination of the motion from dynamical principles will be made much easier than if the two parts were undertaken conjointly.

Here we can see beginnings of the separation of the general problem of dynamics into kinematics and kinetics. Euler's contemporaries I. Kant and D'Alambert (1717–1783) also were treating motion purely geometrically. This is what L.N.M. Carnot (1753–1823) later called "geometric" motion. By the end of the 18th century G. Monge (1746–1818) proposed a course on elements of machines for the École Polytechnique.

Linkages have many different functions.

One of the classical problems in linkage design has been the generation of straight-line paths. Franz Reuleaux (1829–1905), who is often called the "father of modern machine design", had many straight-line mechanisms in his kinematic model collection. Cornell University has a collection with about 220 different F. Reuleaux kinematic models and 39 of them are mechanisms that produce approximate or exact straight-line motion. See the website [29].

In the 20th century, ideas growing from Kempe's work [28] were further generalized by Denis Jordan, Michael Kapovich, Henry King, John Millson, Warren Smith, Marcel Steiner and others. The more recent combinatorial approach to linkages puts together ideas about rigidity, graph theory, and discrete geometry. One can generalize and consider a polygon (where the edges are hinged at the vertices), a tree, or other more general structures consisting of rods and plates (polygons spanned with a rigid membrane). The mathematical study of linkages was an active area of mathematics in the 18th and 19th centuries but then became dormant; however, recently it has been revived by new applications.

Linkages were used during World War II for control and computing. Antonin Svoboda (1907–1980) became involved with computers in 1937 when he started to work for the Czechoslovakian Ministry of National defense. He and Vand designed an antiaircraft gun control system using mechanisms. In 1941 he became a staff member of the Radiation Laboratory working to develop analog computers connected with MARK 56 antiaircraft control. The analog computer had two sections. The linear part, which was called OMAR, was done by linear potentiometers of high precision. But the linear theory had to be modified by nonlinear corrections. Svoboda was asked to produce a mechanical solution for the nonlinear part of the system. See Figure 20 for an example of a linkage computer such as Svoboda designed. Such linkage computers could generate functions of one or two variables. See Svoboda's book [37] for more details.

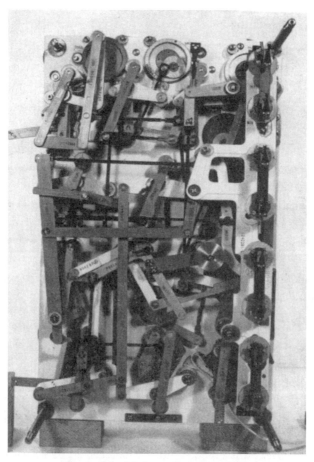

Figure 21. Linkage Computer [23]

For a survey of more recent results on linkages (including some results since 2000), see [6].

Conclusion

In the classroom, drawing of curves first appears in elementary school when students learn how to draw a circle. Often there arises a natural question: Are there instruments that can draw other types of curves? In middle and high school, curves appear as graphs of the functions (for example, quadratic functions for parabolas and reciprocal functions for hyperbolas) and the use of calculators has displaced the actual drawing of these curves — students merely push buttons and the curve magically appears on the screen. Functions and graphing calculators are powerful tools for representing and exploring different curves; however, actual physical construction of a curve is the most valuable step for gaining an initial acquaintance of the curve and exploring its meanings.

It is interesting and important for students, teachers, and mathematicians to explore historical mechanisms that provide connections with the past and that allow the exploration of motions that are related to these mechanisms and that connects the study of curves with mechanical motion design and its applications in engineering. Interactive online resources for the use of mechanisms in classrooms at different levels are being developed and offered by Cornell University's Kinematics Models For Design Digital Library [29], which is built around a large historical collection of mechanisms and includes interactive videos, simulations, tutorials, and on-line historical books.

References

1. Agricola, Georg. *De Re Metallica*. Translated from the first Latin ed. of 1556, with biographical introd., annotations, and appendices upon the development of mining methods, metallurgical processes, geology, mineralogy & mining law from the earliest times to the 16th century, by Herbert Clark Hoover and Lou Henry Hoover, New York, Dover Publications, 1950.

2. Ahrendt M. H., A General Method For The Construction Of A Mechanical Inversor, *Mathematics Teacher*, 37, (1944) 75–80.

3. Artobolevskii I. I., *Mechanisms for the Generation of Plane Curves*, Pergamon Press, New York, 1964.

4. Besant W. H., *Conic Sections Treated Geometrically*, London, 1895,– electronically form Cornell Math Library http://historical.library.cornell.edu/cgi-bin/cul.math/docviewer?did=00630002&seq=11

5. Besson, Jacques Dauphinois. Uniform Title: *Theatrum instrumentorum et machinarum*. French Title: *Theatre des instrumens mathematiques et mechaniques de Iaques Besson ... : auec l'interpretation des figures d'icelui / / par François Beroald ; plus en ceste derniere edition ont esté adioustees additions à chacune figure*. Publisher: A Lyon : Par Iaques Chouët, 1596. Available on-line through [29].

6. Connelly, Robert, Demaine, Eric D., Rote, Gunter; Straightening Polygonal Arcs and Convexifying Polygonal Cycles, *Discrete Comput Geom, 30*: 205–239 (2003).

7. Coolidge, Julian Lowell, *A History of the Conic Sections and Quadric Surfaces*, The Clarendon press, Oxford, 1945.

8. *Ancient Planetary Model Animations*, http://www.csit.fsu.edu/~dduke/models

9. *Dynamic Geometry*, animations by D. Scherr, http://www.edc.org/MLT/DG/

10. Dennis, David, *Historical Perspectives for the Reform of Mathematics Curriculum: Geometric Curve Drawing Devices and Their Role in the Transition to an Algebraic Description of Functions*, PhD Dissertation, Cornell University, May 1995.

11. Descartes, Rene, *La Geometrie,* Paris, 1886. http://historical.library.cornell.edu/cgi-bin/cul.math/docviewer?did=00570001&seq=7

12. Dürer, Albrecht, *Institutionum geometricarum libri quatuor...,* Arnhemiae in Ducatu Geldriae : Ex officina Iohannis Iansonii ..., 1605.

13. Dyck, Walter, *Katalog matematischer und matematisch-physikalischer Modelle, Apparate und Instrumente*, Georg Olms Verlag, Zurich, New York, 1994.

14. Eagles, T. H., *Constructive geometry of plane curves*: With numerous examples, by T. H. Eagles, 1885 In Univ. of Michigan Historical Math collection: http://www.hti.umich.edu/cgi/t/text/text-idx?c=umhistmath;tpl=home.tpl

15. Easton, Joy B., Johan de Witt's Kinematical Constructions of the Conics, *The Mathematics Teacher*, December 1963, pp. 632-635.

16. Euler, Leonard, Novi commentarii Academiae Petrop., vol. XX, 1775; also in *Theoris motus corporum*, 1790. Found in translation in Willis, *Principles of Mechanism*, 2d ed., p. viii, 1870. Available on-line [29].

17. Fergusson, Eugene S., Kinematics of Mechanisms from the Time of Watt, *United States National Museum Bulletin* **228,** Smithsonian Institute, Washington D.C., 1962, pp. 185–230. Available on-line through [29].

18. Field, Peter, *Projective geometry, with applications to engineering*, Van Nostrand Company, New York, 1923 online in University of Michigan Historical Math Collection:

http://www.hti.umich.edu/cgi/t/text/text-idx?sid=8675e87426e349a1d439e95a34ae01fd;c=umhistmath;idno=ACV3228.0001.001

19. Galle A., *Matematische Instrumente*, Teubner Verlag, Leipzig und Berlin, 1912. Online in Cornell Library Historical Math Collection: http://historical.library.cornell.edu/cgi-bin/cul.math/docviewer?did=03650002&seq=7

20. Hartenberg, Richard S., Denavit, Jacques, *Kinematic Synthesis of Linkages*, Mc Graw Hill Book Company, 1964.

21. Henderson, D., Taimina, D. *Experiencing Geometry: Euclidean and Non-Euclidean with History*, 3rd ed., Prentice Hall, Upper Saddle River, NJ, 2005.

22. Hilsenrath, Joseph, Linkages, *Mathematics Teacher*, 30, (1937) 277–284.

23. Hinkle, Roland T., *Kinematics of Machines*, 2nd. Ed., Prentice Hall, Englewood Cliffs, NJ, 1960.

24. Honnecourt, *The sketchbook of Villard de Honnecourt*. Edited by Theodore Bowie. Published. Bloomington, Indiana University; distributed by G. Wittenborn, New York, c1959.

25. Hopcroft, J., Joseph, D., Whitesides, S., Movement problems for 2-Dimensional Linkages, *SIAM J. Comput.* Vol. 13, No.3, August 1984.

26. Horsburgh, E. M., editor, *Modern Instruments and Methods of Calculation, A Handbook of the Napier Tercentenary Exhibition*, G. Bell and Sons and The Royal Society of Edinburgh, London, available online at: http://historical.library.cornell.edu/cgi-bin/cul.math/docviewer?did=03640002&seq=7

27. Kanayma, Raku. Bibliography on the Theory of Linkages, *Tohoku Mathematical Journal*, 37 (1933) 294–319.

28. Kempe, A. B., *How to Draw a Straight Line,* London: Macmillan and Co. 1877. Available on-line through [29].

29. KMODDL (Kinematics Models For Design Digital Library), www.kmoddl.library.cornell.edu

30. Lloyd, Daniel B., The Teaching Of "Flexible" Geometry, *Mathematics teacher*, 32, (1939) 321–323.

31. Mathworld, http://mathworld.wolfram.com/TusiCouple.html

32. McCarthy J. M., *Geometric Design of Linkages*, Springer, New York, 2000.

33. Meserve, Bruce E., Linkages As Visual Aids, *Mathematics Teacher*, 39, (1946) 372–379.

34. Moon, Francis C., tutorial "How to draw an ellipse", see [29].

35. Proclus, *A Commentary on the First Book of Euclid's Elements*. Princeton: Princeton University Press, 1970.

36. Schooten, Frans van, *Mathematische oeffeningen: begrepen in vijf boecken*. Amsterdam: Gerrit van Goedesbergh, 1659–60.

37. Svoboda A., *Computing Mechanisms and Linkages*, Dover, New York, 1965.

38. Taylor, Charles, *An introduction to the ancient and modern geometry of conics, being a geometrical treatise on the conic sections with a collection of problems and historical notes and prolegomena* http://historical.library.cornell.edu/math/math_T.html

39. *Theatrum Machinarum*, stored in the Laboratory of Mathematics of the University Museum of Natural Science and Scientific Instruments of the University of Modena. http://www.museo.unimo.it/theatrum/

40. Yates, Robert C., The Story Of The Parallelogram, *Mathematics Teacher*, 33, (1940) 301–309.

41. Yates, Robert C., *Tools: A Mathematical Sketch and Model Book*, Louisiana State University, 1941.

Thanks to National Science Foundation (research grant DUE-0226238) and Institute of Museum and Library Services (research grant LG-30-04-0204-04) for partial support of the work on this paper.

Learning from the Roman Land Surveyors: A Mathematical Field Exercise

Hugh McCague

York University

Introduction

In the development and rise of civilizations and empires, land surveying has played a major role because it is crucial to the imposition and maintenance of system, order, and control of the landscape through the demarcation of properties, boundaries and roads. The key behind this system and order is always mathematics. We will focus on the Romans who had a highly developed system of land surveying as attested by their surveying manuals, and land divisions, town plans, architecture and engineering works still to be seen throughout the wide expanse of the earlier Roman Empire and Republic [1]. Indeed, some of these features in the landscape are still in use today. Additionally, we are fortunate to have extensive writings of Roman land surveyors, the *Corpus Agrimensorum* [2], describing many aspects of their work and methods. Some archaeological artifacts pertaining to surveying equipment have been found and analyzed. For example, the metal parts of a *groma* surveying instrument were unearthed in the workshop of the surveyor Verus at the ruins of Pompeii in southern Italy. Also, the tombstone of the surveyor Lucius Aebutius Faustus from Ivrea in northern Italy depicts a dismantled *groma* [3]. From these various sources a great deal is known about Roman land surveying and its central use of mathematics and geometry. With this considerable background to draw upon, we will try our hand at Roman land surveying and learn more about mathematics, geometry and history in the process.

What are a *groma* and a *decempeda*?

Two of the most important Roman surveying instruments were the *groma* [Figure 1] for sighting right angles and the *decempeda,* a measuring rod of 10 Roman feet. We will reconstruct simple working models of the *groma* and the *decempeda* measuring rod. The late O. A. W. Dilke, a classicist and scholar on Roman land surveying, developed a plan for constructing the *groma* and *decempeda* measuring rod which we will use and adapt here [4]. I took detailed measurements of the *groma* model at the Science Museum, London, which was based on the *groma* model in the National Museum, Naples. The latter model was based on the archaeological finds of Matteo Della Corte at Pompeii [5].

Based on historic texts [6] and archaeological evidence [7], two different reconstructions of this mathematical instrument, the *groma,* have been proposed [8]. The *groma* consisted of a pole or staff surmounted by an offset arm or bracket upon which an equal-armed cross-piece could rotate. The bottom end of the pole had a pointed shoe which was thrust into the ground, so that the pole was vertical and the cross-piece horizontal. Four plumb-lines were attached from the outer ends of the cross-piece. A straight line could be

Figure 1. Roman *groma*. (Image: Redrawn after O. A. W. Dilke, *Surveying the Roman Way,* University of Leeds, Leeds, 1980.)

surveyed by viewing two plumb-lines opposite each other. Similarly, another straight line could be surveyed by viewing the other two plumb-lines opposite each other. These two straight lines thus formed were at right angles to each other. The bracket allows the plumb-lines to be viewed unobstructed by the vertical staff or pole. An alternative reconstruction of the *groma* is the same as the above, except there is no bracket. The cross-piece is set atop the pole with a connecting metal cylinder or wooden dowel. The pole would have to be tipped slightly off vertical to allow some unobstructed sighting along the plumb-lines. This slight tipping of the pole off vertical would, of course, lead to errors, but those inaccuracies can be negligible [9].

Instructions for the *groma* and *decempeda* construction

The Staff of the *Groma* The full length of the wooden staff with pointed metal shoe should be about 1.80 m or 5′ 11″. The staff can be about 3 cm or 1.25″ in diameter. For the stability of the *groma,* the pointed shoe should be long enough to be able to penetrate the soil to a depth of about 15 cm or 6″. I suggest you use a metal spike with a heavy wood mallet to prepare a hole in the soil for the staff's shoe to enter. Soil, even when quite wet, is hard to penetrate with the staff unless a hole is already prepared. (Image: Redrawn after O. A. W. Dilke, *Surveying the Roman Way,* University of Leeds, Leeds, 1980.)	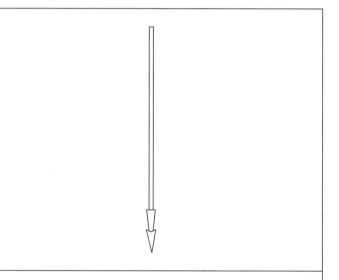
The Bracket of the *Groma* The bracket found at Pompeii had a metal surround and would have had wood inside. Your bracket can be made entirely of wood. Make it about 27 cm or 11″ in length. The lower cylinder A can be about 13.5 cm or 5.5″ in height and about 4.5cm or 2″ in diameter. The upper cylinder B can be about 6 cm or 2.5″ in height and about 3.5 cm or 1.5″ in diameter. You can simplify the design, as long as it operates in the intended manner. (Image: Redrawn after O. A. W. Dilke, *Surveying the Roman Way,* University of Leeds, Leeds, 1980.)	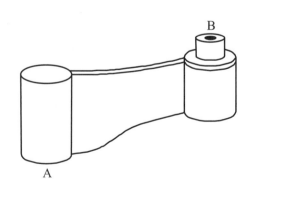

The base or cylinder A needs a drilled hole to allow the top projection of the staff to fit within. The top of cylinder B needs a smaller cylindrical or dowel protrusion in order to receive the cross described below.

An eye-screw can be placed at the center of the bottom of the cylinder B for the attachment of an extra plumb-line. Strictly, there does not appear to be historical or archeological evidence for this extra plumb-line, but its addition is quite practical, so one wonders if some Roman surveyors did likewise.

There is some question about whether the Roman *groma* actually had a bracket. I suspect it did. You can use your *groma* with or without the bracket and see which arrangement seems more practical to you. To use the *groma* without the bracket, use a small wood dowel that fits well with the staff and cross at C, and place the staff slightly off vertical to allow the sighting along the plumb-lines.
(Photograph by H. McCague.)

The Cross of the Groma

The dimensions in centimeters based on the Pompeii *groma* finds are given in the figure opposite. Use 2 wooden arms joined together. The cross design can be simplified so that the cross maintains a constant width. A drilled hole is needed at the center to accept the bracket. The arms of the cross can be kept fixed at right angles by metal angle-brackets.
(Image: Redrawn after O. A. W. Dilke, *Surveying the Roman Way*, University of Leeds, Leeds, 1980.)

A small hole needs to be drilled near the end of each arm to allow the passage of the string for the plumb-line. For the accuracy of the *groma*, take care that the 4 holes and the center of the cross form 4 accurate right angles.

You can cut narrow slits into the top of the cross near the small holes so that the fed strings can be wrapped around the end of the cross and secured (or 'pinched') in the tight fit of the cut slits.
(Image: Redrawn after O. A. W. Dilke, *Surveying the Roman Way*, University of Leeds, Leeds, 1980.)

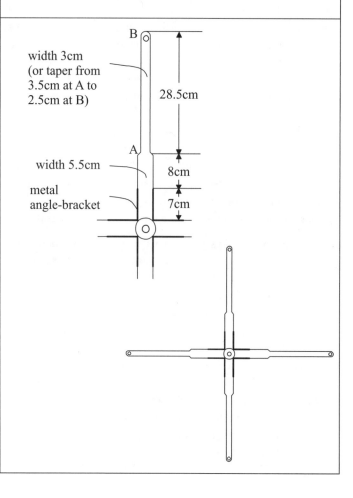

Plumb-Lines of the *Groma*

Attach plumb bobs to the ends of string or narrow cords about 1m or 1 yard in length. To economize, heavy nuts and bolts can be used instead of plumb bobs. I suggest that one pair of matching plumb bobs (or bolts) be used diametrically opposite each other, and similarly, a different pair of matching plumb bobs (or nuts) be used opposite each other.

(Image: Redrawn after O. A. W. Dilke, *Surveying the Roman Way*, University of Leeds, Leeds, 1980.)

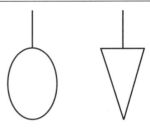

As mentioned in the description of the bracket, a hypothetical fifth plumb-line could be hung from underneath the bracket to help insure the *groma* is properly centered over a wooden stake which is the survey station point.

(Photograph by H. McCague.)

Wooden stakes and an offset stick

Wooden stakes with a square cross-section and ends cut to a stake are needed to mark the vertices (the survey stations) of the surveyed rectangles. The number of stakes needed will depend on how involved a survey project you undertake. However, I suggest you make at least 16 stakes for your first Roman surveying exercise. For the accuracy of your survey work, draw the diagonals of the square with pencil on top of the stake. The intersection of the diagonals will indicate the center of the stake and a vertex (a survey station) of the rectangles and squares you will be surveying. If a fifth central plumb-line is used, this plumb bob needs to be right over the center of the stake [Figure 4]. A wooden mallet, mentioned in regard to the staff of the *groma,* is well suited for pounding the stakes into the soil.

In addition to the stakes, you may wish to make a wooden offset stick sized to provide exactly the same offset as the bracket near the top of the *groma*. One end of the offset stick can be cut with a semi-circular 'arch' to fit snugly with the staff of the *groma*. The other end of the offset stick can be cut to a broad 'point.' Draw a central axis line on the top of the offset stick. Along with a fifth central plumb-line described above, this device can be used to help place the *groma* accurately over the center of the wooden stake (survey station) below. Strictly, there does not appear to be historical or archeological evidence for an offset stick, but its addition is quite practical, so again one wonders if some Roman surveyors did likewise.

(Photograph by H. McCague.)

Learning from the Roman Land Surveyors: A Mathematical Field Exercise

Two Measuring Rods

Make two measuring rods of 10 Roman feet. Use the common Roman foot unit of the *pes Monetalis* equal to approximately 29.57 cm, or about 11 5/8" (as opposed to the British Imperial foot, 30.48 cm, 12", used elsewhere in this article). Thus, the rods need to be about 295.7 cm or about 9' 8 7/16". I suggest two such rods, so that they can be placed end to end to facilitate the measuring out process.

(Image: Metal end redrawn after O. A. W. Dilke, *Surveying the Roman Way*, University of Leeds, Leeds, 1980. Photograph by H. McCague)

The metal ends as pictured here are based on an archaeological find near Enns in northern Austria [10]. If possible, make your ends from metal as well and about 10cm or 4" long. A wooden dowel about 4 cm or 1.5" in diameter should fit inside. The hard metal ends protected the softer wooden ends from damage and shortening due to 'wear and tear' over years of use in the field.

However, if you wish to simplify, you could use the dowels without the metal ends, but still take care to measure and cut their length accurately. To ease their storage and transport to site, you may wish to have the rods in two halves with a metal sheath to hold them together when in use. However the rods are made, they need to be stiff for accurate usage.

Roman land measurements

Soon we can try our hand at surveying in the Roman fashion. First, however, we will need to learn some more about Roman surveying theory and practice. In practice, Roman surveying tools including the *groma* and *decempeda* were used to survey and lay out military camps, forts, towns, and large field and road systems.

The *actus* of 120 Roman feet was the fundamental unit in Roman land surveying. As noted earlier the most common Roman foot unit was the *pes monetalis* (at or very close to 29.57cm or 11 5/8"). The basic unit of area was the *iugerum* equal to a rectangle of 2 *actus* by 1 *actus*. 2 *iugera* equal 1 *heredium* (2 *actus* by 2 *actus* or 240 Roman feet by 240 Roman feet) [Figure 2]. 100 *heredia* together make a *centuria* (20 *actus* by 20 *actus* or 2400 Roman feet by 2400 Roman feet) [11].

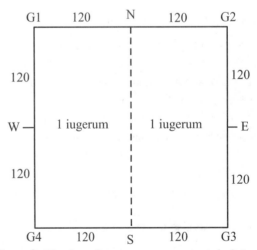

Figure 2. The *heredium* area measure equals 2 *iugera*. The 120 Roman feet marked equals an *actus*. (Redrawn after O. A. W. Dilke, *Surveying the Roman Way*, University of Leeds, Leeds, 1980.)

In large colonies, many *centuriae* were surveyed together, somewhat like a chessboard except with boundary roads or tracks the Romans called *limites* [Figure 3]. The survey of a colony would begin with the two major *limites* road axes oriented and at right angles to each other: the *decumanus maximus* (DM) and the *cardo (kardo) maximus* (KM). DM and KM are E-W and N-S axes, respectively, though, in practice, some variations from those alignments occurred. A labeling system, somewhat resembling Cartesian co-ordinates [12], with Roman numerals then designated the location of the *centuria*. Note the following Roman terminology in Figure 3:

S.D. *sinistra decumani*   to the left of *decumanus*
D.D. *dextra decumani*   to the right of *decumanus*
C.K. *citra kardinem*   to the near side of *kardo*
V.K. *ultra kardinem*   beyond *kardo* [13].

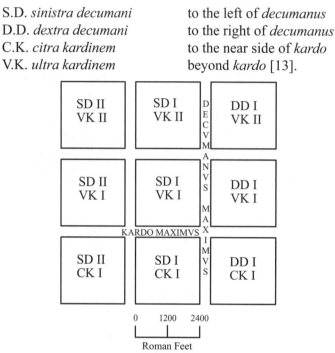

Figure 3. *Centuriae.* Centuration was the Roman method of land apportionment based on their linear and area standard measurement systems. (Redrawn after O. A. W. Dilke, *Surveying the Roman Way,* University of Leeds, Leeds, 1980.)

Using your *groma* and two *decempedae*

A description is given below for the use of the *groma* with the bracket based on a surveying exercise by O. A. W. Dilke [14]. The instructions can be used with the *groma* without the bracket, but the wooden stakes will need to be placed and replaced in the same holes used by the pointed shoe. Windy conditions can overly disturb the use of the plumb-lines. Also, avoid rainy conditions that can cause the wood of the *groma* to swell and even the bracket to get stuck and unable to swivel in the staff. The latter state of affairs may have been mitigated in the actual Roman *groma* through protective metal coverings as found at Pompeii. It is helpful to have two or more people for these Roman surveying exercises. A large football or soccer field is suitable for the survey site. An open park area or school ground that is fairly level with low grass can also be used. We will survey a *heredium* which equals 2 *iugera* [Figure 2].

Surveying exercise instructions

1. In a corner of a large field, set the assembled *groma* vertically over a stake (survey station) [Figure 4] to mark point or vertex G1 [Figure 2]. Use the plumb-lines to check the staff for vertical each time the *groma* is set up.
2. Temporarily, set a second stake along the line segment G1G2, say 30 steps away.

Figure 4. Laying out a survey design with a re-created Roman groma and measuring rod. Note that the fifth central plumb-line is right over the center of the wooden stake. The stake marks a survey station point which is a vertex of a rectangle being surveyed. (Photograph by H. McCague.)

3. Rotate the cross so that one of the pairs of opposite plumb-lines align exactly with the distant stake along G1G2. Make sure that the *groma* and bracket are not moved.

4. Use the two *decempedae* to measure 240 Roman feet *(pes monetalis)* along G1G2 as sighted in Step 3 above. One person can remain sighting at the *groma* while one or two persons alternate measuring with the two rods end to end on the line segment G1G2. Set a stake at G2.

5. Now sight along the other pair of plumb-lines, and similarly measure out 240 Roman feet *(pes monetalis)* to establish G4 with a stake.

6. Move and center the *groma* over G2. Rotate the cross so that a pair of opposite plumb-lines align with G1. Again, make sure that the *groma* and bracket are not moved.

7. As before, using the alignment sighted by the other pair of plumb-lines measure out 240 Roman feet *(pes monetalis)* to establish G3 with a stake.

8. Move and center the *groma* over G3. Rotate the cross so that a pair of opposite plumb-lines align with G2. Make sure that the *groma* and bracket are not moved. Now check to see if the other pair of plumb-lines align with G4. Also, check if G3G4 is 240 Roman feet *(pes monetalis)* as intended. Sufficient inaccuracies require one to retrace, and redo earlier steps where necessary.

9. Similar to Step 8, one can move and center the *groma* over G4 and check the right angle. Again, sufficient inaccuracies require one to retrace, and redo earlier steps.

10. The diagonals of the large square *heredium* are then measured as a check on the right angles and side lengths. By applying the Pythagorean Theorem, both the diagonals need to be close to $240\sqrt{2} \approx 339.4 \approx 340$ Roman feet *(pes monetalis)*. 34/24 or, more simply, 17/12 is a close rational approximation to the irrational $\sqrt{2}$. The Roman surveyors and builders understood this matter in geometric terms: 34:24 can be taken as 'equal' to, or more correctly is a close approximation to, a square's diagonal to its side [15].

If you have the opportunity to use some modern surveying equipment, or even a long and accurate tape measure, you can do a modern check of the accuracy. As noted already, you can apply the Pythagorean Theorem to check the length of the two measured diagonals of the surveyed square, and hence check if right angles were achieved. As an additional check on your measurements and calculations, you can calculate the angles of the surveyed 'square' using trigonometry, more specifically the Cosine Law. Additionally, the surveyed angles, intended to be right angles, can be directly measured and checked by using a surveyor's transit or theodolite. With care in construction and use of the *groma*, the surveyed angles can all be within 1/2 of a degree (30 minutes), or even 1/6 of a degree (10 minutes), of 90 degrees.

Once you have obtained some practice at the above basic exercise, you may wish to try laying out a larger or more involved survey including more squares and even some *limites* or dividing tracks similar to Figure 3. If the field in use is not big enough, you can make the sides of the squares shorter, for example 60 or 120 Roman feet *(pes monetalis)*.

Further mathematical exploration of the groma and Roman land surveying

There are many more mathematical issues on the topic of Roman land surveying that you may enjoy learning about. Even though the *groma* was used to simply sight straight lines and right angles, it, along with other Roman surveying instruments and methods, nevertheless offers a wide range of geometric and trigonometric issues of historical and educational note [16]. Additionally, the collection of Roman land surveying manuals, the *Corpus Agrimensorum,* includes some geometric techniques and methods, without proofs, that show the use of the *groma* to sight straight lines not part of right angles. It also includes associated applications of basic laws of triangles, parallelograms, and even 3-dimensional geometry [17]. For example, Hyginus Gromaticus, *Establishment of Boundaries,* 155.17-156.15, in effect, applies the laws of similar triangles and parallelograms to show how a line segment can be surveyed that is parallel to another line segment determined by two inaccessible points. M. Junius Nipsus, *Measuring the Width of a River,* 4-28, uses congruent triangles to determine the width of a river by surveying the land on one side of the river [18]. The Roman surveyors commonly used two shadows cast from a vertical rod, called a gnomon, during the morning and at a certain corresponding time during the afternoon to determine the four cardinal directions [19]. You may wish to explore how this method works. Intriguingly, the path of the tip or apex of the shadow over the course of the day is approximates very closely a conic section based on the latitude and sun's declination [20]. In addition, exploring the small inaccuracies from a right angle caused by a slightly tilted *groma* without a bracket is a good exercise in 3-dimensional geometry and trigonometry [21].

The significance of the *groma*

The *groma,* standard measurement units and land surveying were critical to the expansion and maintenance of the large Roman Empire and Republic. The use of the *groma* did not apparently continue during the medieval period [22]. Of significant note in the history of mathematics, the *Corpus Agrimensorum* was formative in the later developments of practical geometry during the Middle Ages, particularly through such eminent scholars as Gerbert and Hugh of St. Victor [23]. Also, today's sophisticated surveying technology and practices have their early origins in the still remarkable achievements and applications of mathematics of the Romans and other ancients.

Further, for the Romans, the geometry applied in the design and use of the *groma* was deemed to have a heavenly origin and was patterned after the divine archetype of the *templum* of the sky, a celestial temple pattern. One of the writers of the *Corpus Agrimensorum,* copied during the Middle Ages, states that centuration had heavenly origins, and the Roman author Varro mentions that it was adopted from Etruscan ritual [24]. The cross-piece of the *groma* and the surveyed cross (of right angles) on earth manifested the cross of the heavenly *templum,* said to have descended beforehand during augural ritual. At the founding of a new settlement, the Roman augur would inwardly see the *templum* descend from the sky and heavens to form and make sacred and habitable the selected site. This preparation was deemed necessary to commence the surveying of the land [25]. In the ancient world, mathematics and technology were intimately mingled with philosophy, religion and the purpose of life, and Roman land surveying was no exception. In a similar spirit, this field exercise with the *groma* and *decempeda* is intended to increase our appreciation of the application of mathematics and its great pertinence to philosophy, history and culture.

References

1. For the reader interested in gaining both a broader and more detailed understanding of Roman land surveying and its context, some books are listed here. O. A. W. Dilke, *Mathematics and Measurement,* British Museum Publications, London, 1987, provides a concise overview of the history of measurement units, instruments for measuring and surveying, and related mathematics in ancient Egypt, Mesopotamia, Greece, and Rome. ——, The *Roman Land Surveyors: An Introduction to the Agrimensores,* David and Charles, Newton Abbot, Devon, England, 1971, is an excellent and broad study of the Roman land surveyors. Jean-Pierre Adam, Roman *Building: Materials and Techniques,* translated from the French by Anthony Mathews, Indiana University Press, Bloomington and Indianapolis, 1994, discusses and illustrates Roman land surveying and the broader subject of Roman building construction. M. J. T. Lewis, *Surveying instruments of Greece and Rome,* Cambridge University Press, Cambridge and New York, 2001, provides a more in-depth and up-to-date study of Greek and Roman surveying instruments.

2. J. B. Campbell, *The writings of the Roman land surveyors: introduction, text, translation and commentary,* Journal *of Roman Studies* monograph, no.9, London, Society for the Promotion of Roman Studies, 2000.

3. Dilke, *The Roman Land Surveyors,* 39, 50.

4. O. A. W. Dilke, *Surveying the Roman Way,* University of Leeds, Leeds, 1980. (This document is an instructions kit. It is not generally available.) I would like to acknowledge the assistance of the late Len Patrick in the construction of my *groma* and two *decempedae*.

5. Matteo Della Corte, "Groma," *Monumenti Antichi,* 28 (1922) 5–100. Dilke, *The Roman Land Surveyors,* 50, 69–70.

6. Lewis, *Surveying instruments of Greece and Rome*. Note particularly the ancient descriptions of the *groma*: 309–310, Source 14; 324–325 Source 51; 328 Sources 56 and 57. References to the *groma* can also be checked in Campbell, *The writings of the Roman land surveyors*.

7. Dilke, The *Roman Land Surveyors.* Thorkild Schiöler, "The Pompeii-groma in New Light," *Analecta Romana Instituti Danici,* XXII (1994) 45–60.

8. Schiöler, "The Pompeii-groma in New Light."

9. Schiöler, "The Pompeii-groma in New Light," 55. The geometry and trigonometry associated with the resulting errors are interesting to explore.

10. Dilke, The *Roman Land Surveyors*, 67.

11. Dilke, *Surveying the Roman Way.* ——, The *Roman Land Surveyors.* ——, *Mathematics and Measurement.* Note particularly Chapter 5: "Mathematics for the Surveyor and Architect." A *centuria* of 20 *actus* by 20 *actus* appears to have been standard during the Late Roman Republic and the Roman Empire, but other sizes did occur. John Peterson, "Centuration" entry in "A glossary of terms used in Roman land surveying," "Ancient Landscapes, Information Systems and Computers" [updated 4 April 1996; accessed 13 September 2006]. Available from http://www.sys.uea.ac.uk/Research/researchareas/JWMP/glossary.html.

12. More generally, planar co-ordinate systems predate the work of René Descartes on analytic geometry, as noted by Carl B. Boyer, "The Invention of Analytic Geometry," *Scientific American,* 180.1 (1949): 40–45. Also, note Julian Lowell Coolidge, *A History of Geometrical Methods,* Clarendon Press, Oxford, 1940, 117ff.

13. Dilke, *Mathematics and Measurement,* 31.

14. Dilke, *Surveying the Roman Way.*

15. For example, the later Roman building manual by Marcus Cetius Faventinus mentions that a builder's square can be made to professional standards by joining, in the form of a triangle, three scales with lengths respectively 2

feet, 2 feet, and 2 feet 10 inches (i.e. 24″, 24″, and 34″). This statement is reasonable because 34/24 is, as we have noted, a close approximation to √2. Hugh Plommer, *Vitruvius and Later Roman Building Manuals* including the Latin text and translation of Marcus Cetius Faventinus, *De diversis fabricis architectonicae*, Cambridge University Press, London, 1973. Note *De Diversis Fabricis Architectonicae,* 28, in Plommer, 81.

16. Dilke, *Mathematics and Measurement*. Note particularly Chapter 5: "Mathematics for the Surveyor and Architect." Monique Clavel-Leveque, "Centuration, géométrie et harmonie: Le cas du biterrois" in *Mathématiques dans l'Antiquité,* Jean-Yves Guillaumin, ed., Centre Jean-Palerne, Mémoires XI, Publications de l'Université de Saint-Étienne, Saint-Étienne, 1992, 161–184. John Peterson, "Trigonometry in Roman cadastres" in *Mathématiques dans l'Antiquité,* Jean-Yves Guillaumin, ed., Centre Jean-Palerne, Mémoires XI, Publications de l'Université de Saint-Étienne, Saint-Étienne, 1992, 185–203. More relevant articles by Peterson appear on his web site ——, "Ancient Landscapes, Information Systems and Computers" [updated 3 April 2005; accessed 13 September 2006]. Available from http://www.sys.uea.ac.uk/Research/researchareas/JWMP/.

17. Karl Röttel, "Aus der Arbeit der römischen Feldmesser," *Praxis der Mathematik,* 23.7 (1981) 210–215. O. A. W Dilke, The *Roman Land Surveyors*. ——, "Illustrations from Roman surveyors' manuals," *Imago Mundi,* XXI (1967): 9–29. Note the mathematical proof associated with the more involved and less practical 3-shadow method for determining the 4 cardinal directions, pp. 17–18, Plate 6, Fig.2f.

18. Lewis, *Surveying instruments of Greece and Rome,* 326–327, Sources 54–55. Dilke, *The Roman Land Surveyors*, 59, 61.

19. *De munitionibus castrorum,* c.100 AD, a Roman camp-making treatise by Pseudo-Hyginus Gromaticus, gives the 2-shadow method, quoted in translation from the Latin in in Dilke, *The Roman Land Surveyors,* 57–58. Note the manuscript illustration reproduced in Dilke, "Illustrations from Roman surveyors' manuals," Plate 6 Fig.2d, p.17.

20. Heinrich Dörrie, *100 Great Problems of Elementary Mathematics: Their History and Solution,* translated from the German by David Antin, Dover Publications, New York, 1965, 340–342. This reference assumes that the sun's declination remains the same over the course of a day for a given latitude. This assumption is, of course, not strictly correct, but the errors involved are very small.

21. Schiöler, "The Pompeii-groma in New Light," 55. Note also the relevant text and 3-dimensional geometric analysis of Hero (Heron) of Alexandria, *Dioptra,* 33 translated in Lewis, *Surveying instruments of Greece and Rome,* 282–283.

22. Lewis, *Surveying instruments of Greece and Rome*. David Friedman, *Florentine New Towns: Urban Design in the Late Middle Ages,* Architectural History Foundation, New York; MIT Press, Cambridge, MA and London, 1988. 256–258 n.33. Friedman gives a comprehensive and concise exposition of medieval land surveying in the West.

23. Lon R. Shelby, "The Geometrical Knowledge of Mediaeval Master Masons," *Speculum,* XLVII (1972) 395–421. Frederick A. Homann in Hugh of Saint-Victor, *Practical geometry (Practica geometriae) attributed to Hugh of St. Victor*, translated from the Latin with notes by Frederick A. Homann, Marquette University Press, Milwaukee, 1991.

24. O. A. W. Dilke, "Varro and the Origins of Centuration" in *Atti del Congresso di Studi Varroniani.* Rieti: 1976. II: 353–358. ——, "Religious Mystique in the Training of Agrimensores" in *Res Sacrae: Homages à Henri Le Bonniec,* D. Porte and J.-P. Néraudau, eds., Collection Latomus, 201, Latomus, Revue d'Études Latines, Brussels, 1988. 158–162, especially 158–159.

25. Joseph Rykwert, *The Idea of a Town: The Anthropology of Urban Form in Rome, Italy and the Ancient World,* Princeton University Press, Princeton, 1976. 45ff.; Paul Godfrey and David Hemsoll, "The Pantheon: temple or rotunda?" in *Pagan Gods and Shrines of the Roman Empire,* Monograph No. 8, Oxford University Committee for Archaeology, Oxford, 1986. 195–209, especially 200–201.

Equating the Sun: Geometry, Models, and Practical Computing in Greek Astronomy

James Evans
University of Puget Sound

Introduction

Ancient Greek planetary theory was geometrical in spirit. Each planet was deemed to ride on a circle, or combination of circles. Greek planetary theory, conceived in this way, originated late in the third century B.C. with the work of Apollonios of Perga. At first, such theories offered only a broad explanation of planetary phenomena: each planet generally travels eastward around the zodiac, but occasionally reverses direction and travels in retrograde motion toward the west for a few weeks or months (depending on the planet), before reverting to its normal eastward motion. Apollonius's geometrical models were under-girded by Aristotle's philosophy of nature, which postulated the centrality of the Earth and the primacy of circular motion for celestial bodies.

In the middle of the second century B.C., Hipparchos, who had been strongly influenced by his contact with quantitative Babylonian astronomy, insisted on the importance of having geometrical theories that also reproduce the phenomena in a *quantitative* sense. But Hipparchos was able to achieve this only for the Sun and the Moon. The Greek theory of the planets was brought into its final, very successful form by Ptolemy only in the second century A.D. Now it became possible to calculate positions for the planets (as well as the Sun and Moon) from geometrical models based on circular motion. Thus, by Ptolemy's day, Greek planetary theory offered not merely a broad, explanatory view of how the universe works: it also provided a tool for practical, predictive calculation of planetary phenomena. Moreover, Ptolemy's planetary theories are usually pretty accurate. Because the theories were in accord with prevailing ideas of Aristotelian physics (or philosophy of nature), the complete package is very imposing. This explains why it lasted for nearly a millennium and a half. Everything fit: astronomy, geometry, and ancient physics formed a unified whole. And it really worked! [1]

Modern readers are often uncomfortable with a theory that places the Earth (rather than the Sun) at rest in the center of things. But the choice of center is a free one, for it is merely the choice of a frame of reference, and we can always ask what the universe would look like from the Earth. The answer, of course, is that, viewed from the Earth, the universe looks more or less like Ptolemy's description of it. A Sun-centered planetary theory is not automatically any more accurate than an Earth-centered one. This is one reason why astronomers were able to work with perfect contentment using Earth-centered theories for the better part of two thousand years. If it were easy to tell what is really moving and what is really at rest, we would not have had to wait until the astronomical revolution of the sixteenth century for the big switch. In all that follows, we will take the ancient point of view, that the Earth is at rest at the center of things.

Because of its geometrical character, Greek astronomy lent itself readily to models and instruments of various sorts. Celestial globes, figured with the constellations, could be used to demonstrate not only the daily rotation of the universe, but also various propositions of spherical astronomy. More complicated mechanisms (today called orreries), which were worked by gears and sometimes run by water power, simulated the intricate dance of the planets. Although the orreries were rare, the globes were quite common and played a vital role in astronomy instruction, as we can tell from references to them by Strabo [2] and Geminos. [3]

More significantly, specialized calculating devices called *equatoria* were also constructed. An equatorium not only provides a visual display of planetary phenomena, but also allows the user to obtain quantitative predictions (as well as retrodictions) of planet positions, while avoiding most of the calculation. With the aid of an equatorium, the user can obtain the longitude in the zodiac of the Sun, Moon and planets in a matter of minutes, rather than the hours that might be required for a straightforward trigonometric calculation from Ptolemy's theory, or even using Ptolemy's tables in the *Almagest*.

In this chapter, I will explain what an equatorium does and show how it works, taking the simplest case, that of the Sun, for the example. I also provide directions for making and using a solar equatorium, based on a Renaissance model. If we are to make sense of the equatorium, and appreciate the great simplicity it affords in practical calculation, we must first understand something about the ancient Greek solar theory. That is, we can't appreciate the beauty of the solution offered by the equatorium, until we understand the nature of the problem it was intended to solve, as well as what the alternative (trigonometric) solutions of the same problem look like. So we must begin with an overview of the ancient solar theory. Along the way I will also pose two practical astronomy problems that mathematics teachers can use with students who have had a bit of trigonometry.

Hipparchos's theory for the motion of the sun

In a simple, preliminary model, we can represent the Sun as (1) moving at constant speed (2) on a circle (3) that is centered on the Earth. Indeed, this is a reasonably good approximation to the facts. But if this model were strictly true, the equinoxes and solstices, which are evenly spaced around the zodiac, would be evenly spaced in time. If you simply count up the lengths of the four seasons on a modern calendar that is marked with the dates of the equinoxes and solstices, you will find that the longest season (summer in the northern hemisphere) exceeds the shortest (winter) by about 4 days. So the Sun *appears* to move more quickly in some quadrants of the zodiac than in others. One or more of our three assumptions must therefore be given up.

In ancient physical thought (as elaborated in the works of Aristotle, for example), celestial objects, including the Sun, are made of a fifth element, the ether, which is utterly different from the four terrestrial elements of earth, water, air and fire, that make up all things here below. The essential characteristic of the ether is its changelessness (the heavens have always been the same) and its property of moving in uniform circular motion (which is the obvious motion of the stars). For an ancient Greek, therefore, it would have been painful to give up either the uniformity of the Sun's motion or its circular course.

The solution to the dilemma of the Sun's anomalous motion was apparently due to Hipparchos (second century B.C.), who proposed the simple expedient of letting the Sun travel uniformly on a circle that is eccentric to the Earth. It is possible that Apollonios had earlier proposed such a model. But Hipparchos was the first, as far as we know, to use numerical values for the lengths of the seasons to deduce two important parameters of the solar theory: the amount and the direction of the eccentricity. So the first problem to be solved in establishing a working theory of the Sun's motion is illustrated in Fig. 1.

The Earth is at *O* (for "observer"). Lines of sight through *O* connect the two equinoxes (*VE* = vernal or spring equinox, *AE* = autumnal equinox) and the two solstices (*SS* = summer solstice, *WS* = winter sol-

Equating the Sun: Geometry, Models, and Practical Computing in Greek Astronomy

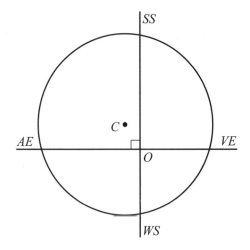

Figure 1. In the solar theory of Hipparchos, the center C of the Sun's circle must be placed eccentrically to the earth O in order to explain the inequality in the lengths of the seasons.

stice). The equinoxes and solstices are equally spaced at 90° intervals as viewed from O. The Sun travels at a uniform speed around the Earth in the course of a year on a circle that is centered at C. The problem then, in the solar theory of Hipparchos, is to determine the position of C so that the lengths of the four seasons agree with the facts of observation:

Lengths of the Seasons (Modern)

	days	hours
Spring	92	19
Summer	93	15
Fall	89	20
Winter	89	0
Total	365	6

In Fig. 2 we have added a line through points O and C. This line cuts the Sun's circle at A the apogee (the place where the Sun is farthest from the Earth in the course of the year) and at perigee Π (where the Sun is closest to us). Astronomers measure celestial longitudes around the ecliptic, starting from the vernal equinox as zero-point. So, the angle marked A is called the "longitude of the Sun's apogee". The distance between C and O is a measure of the off-centerednesss of the circle. But, the only thing that really matters is the ratio of the distance OC to the radius r of the circle. Thus we define the "eccentricity" e of the circle by $e = OC/r$. The eccentricity is therefore a dimensionless number between 0 and 1. So, in somewhat more sophisticated language, our problem is to use the lengths of the seasons to determine the eccentricity of the Sun's circle and the longitude of its apogee.

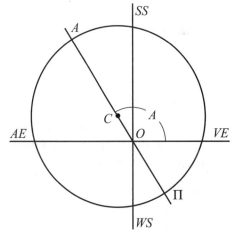

Figure 2. The apogee A and perigee Π of the Sun lie on line OC extended.

Numerical elements for the theory of the sun

This is quite a nice problem for students in a trigonometry class. A beginning of the solution is illustrated in Fig. 3. $EFGH$ is the zodiac in the effectively infinitely distant sphere of the fixed stars. E represents the vernal equinox; *F*, the summer solstice; *G*, the autumnal equinox; *H*, the winter solstice. Points *O* and *C* mark the Earth and the center of the Sun's circle, as before. When the Sun is at *N* on its orbital circle, we see it against *E* and we say that the vernal equinox is occuring. Similarly, at the time of summer solstice, the Sun is at *K* and we see it against *F*. Now add line *QS*, drawn through *C* parallel to *GE*. Similarly, add line *PR*, drawn through *C* parallel to *FH*.

Note that arcs *SKP*, *PAQ*, *QLR* and *RNS* are each 90°.

It is convenient to express the lengths of the seasons in terms of degrees rather than days. Thus, Spring = 92.79 days × (360°/365.25 days) = 91.46°. The other seasons can be similarly expressed.

Then the total length of arc *NKL* is given, as the sum of the spring and summer arcs.

And *QL* = *NS* = (*NKL* – 180°)/2.

It is also easy to find the arcs *PK* and *RM*, which are equal to one another. Note that arc *KPL* is just the length of the summer, expressed in degrees. Then *PK* = *KPL* – *PAQ* – *QL*. But *PAQ* = 90° and *QL* is already known, so *PK* is determined.

With arc *PK* known it is easy to determine the short line segment *CT*. (We may take the radius of the circle to be unity, since the absolute distance of the Sun is irrelevant in working out a theory of the Sun's angular motion.)

In the same way, from arc *QL* we may obtain line segment *CU*.

And then it is any easy matter to do a little trigonometry to find distance *OC* (the eccentricity) and angle *TOC*.

Final answers: from the data for the lengths of the seasons, the eccentricity *OC* = 0.0334, and the longitude of the apogee (angle *NOA*) = 102.4°.

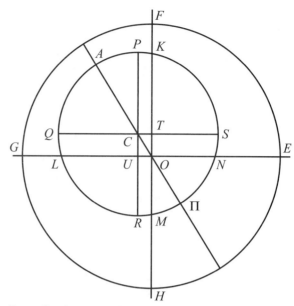

Figure 3. The Sun's circle is centered at **C** and lies eccentrically within the sphere of the cosmos, which is centered at the Earth *O*.

Calculating the sun

Once one has a complete set of elements, one can use the solar theory to calculate the Sun's longitude (its place in the zodiac) for any derived moment.

There are four elements:

1) the mean daily motion *n*, i.e, the number of degrees per day that the Sun moves on its eccentric circle. This is just 360 divided by the length of the year: 360°/365.2422 days = 0.985647°/day.

2) the longitude *A* of the apogee.

3) the eccentricity *e* of the Sun's circle. (We have just seen how to obtain *A* and *e*.)

4) the starting position of the Sun for some one particular moment. For example, suppose that we know the date and time when the Sun last passed through the apogee of its circle.

Then the problem of calculating the Sun's longitude for a general date is illustrated in Fig. 4. Let us define an important quantity: The mean anomaly $\bar{\alpha}$ is the Sun's angular distance past the apogee, as viewed from the center *C* of the Sun's circle. According to Hipparchos's theory, $\bar{\alpha}$ increases uniformly with time.

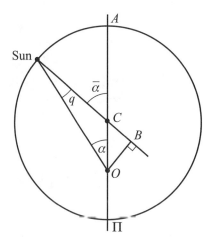

Figure 4. The Sun's mean anomaly $\bar{\alpha}$ increases uniformly with time. The true anomaly α, an angle viewed from the Earth O, therefore increases nonuniformly, because of the off-centeredness of the circle. The equation of center q is the difference in the angular positions of the Sun as viewed by an observer at O and an imaginary observer at C.

(This is another way of saying that the Sun's motion is uniform as viewed from the center C of its circle.) So if we know the time t elapsed since the Sun passed through apogee A, the current mean anomaly of the Sun is just

$$\bar{\alpha} = nt. \tag{1}$$

However, what we need to know is the true anomaly α, i.e., the angular distance of the Sun past apogee, *as viewed from the Earth O*. The relation between α and $\bar{\alpha}$ is

$$\alpha = \bar{\alpha} - q, \tag{2}$$

where q is called the *equation of center*. [Note on terminology: in astronomy, an "equation" is the difference between some nonuniformly increasing quantity and the value this quantity would have if it increased at a uniform rate. Thus the Sun's equation of center is the difference between α (the Sun's actual angular distance beyond apogee) and $\bar{\alpha}$ (the angular distance at which the Sun would be found beyond the apogee if the Sun moved at a uniform angular speed.) It is an equation *of center* because it arises from the fact that the Earth is not located at the center of the Sun's circle.]

It is another nice trigonometry problem to start from Fig. 4 and show that

$$\sin q = \frac{e \sin \bar{\alpha}}{\sqrt{1 + 2e \cos \bar{\alpha} + e^2}}. \tag{3}$$

The maximum value of the Sun's equation of center is about 1.9°, which occurs approximately (though not exactly) when $\bar{\alpha} = \pm 90°$. This means that if we ignore the eccentricity of the Sun's circle and compute the Sun's longitude as if it moved at uniform angular speed around the Earth, we could be off, at times, by as much as 1.9°.

So, to solve for the actual longitude λ of the Sun (its angular distance from the vernal equinox, as viewed from the Earth) at any moment, we would perform the following sequence of calculations: calculate $\bar{\alpha}$ from Eq. (1), calculate q from Eq. (3), calculate α from Eq. (2), then finally, the actual longitude λ of the Sun is given by

$$\lambda = \alpha + A. \tag{4}$$

The solar equatorium

This is a fairly simple sequence of operations, but certainly tedious to perform very often. Fortunately, if we are content with an accuracy of, say, half a degree or so, we can dispense with any need for calculations by

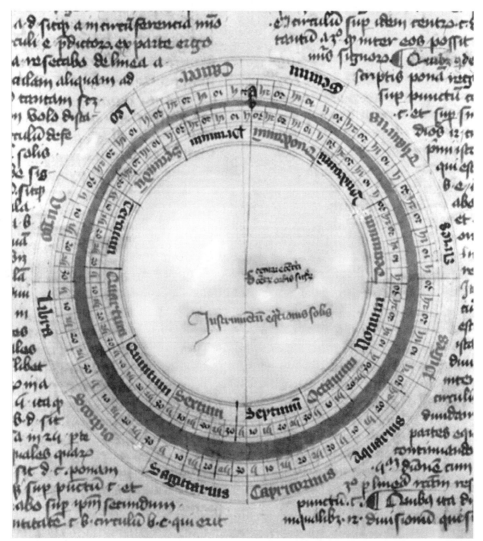

Figure 5. A solar equatorium in an early fifteenth-century manuscript of the *Theorica planetarum* of Campanus of Novara. Photo courtesy of Pepys Library, Magdalene College, Cambridge. Ms. Pepys 2329, fol. 56v.

building a concrete model that incorporates the elements of the solar theory. Such a model is a sort of analog computer that solves the specialized problem of determining the longitude of the Sun for any desired date. In the Middle Ages, such a model was called an *equatorium*, because it supplies the "equation" (of center), which we have denoted q.

The oldest extant equatoria are medieval. But it is likely that the ancient Greeks made them as well. In the introduction to his *Handy Tables*, Ptolemy remarks that one can "calculate" the positions of the planets according to the theories of the *Almagest* by drawing scale diagrams. [4] And it is but a short way from a scale drawing to a reusable instrument. For the case of the Sun, we actually have ancient directions for making an equatorium. Proklos (fifth century A.D.), in his *Sketch of Astronomical Hypotheses* says that on a wooden board or bronze plate, one is to draw a zodiac circle and, within it, an eccentric circle with the proper eccentricity, then divide both into degrees, etc. [5] Proklos is describing an instrument virtually identical to the later one shown in Fig. 5.

Indeed, Fig. 5. shows a beautiful example of a medieval solar equatorium. This comes from a manuscript, dated A.D. 1407, of the *Theories of the Planets*, written by Campanus of Novara in the thirteenth century. This was one of the first competent expositions of Ptolemaic planetary theory to be written after

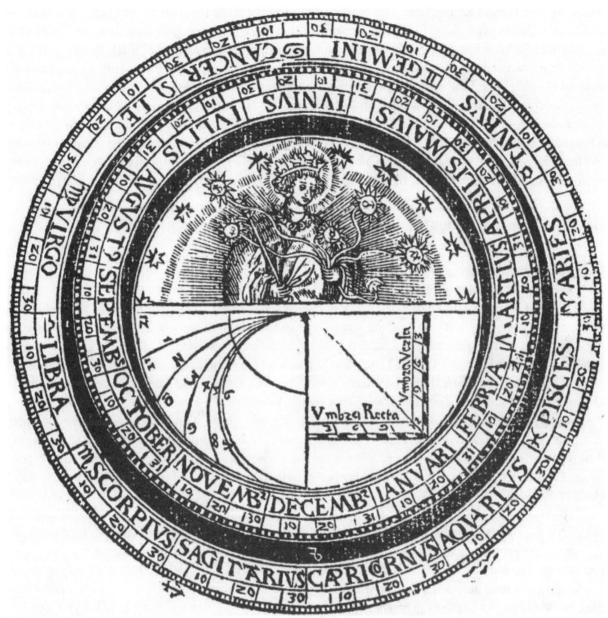

Figure 6. A solar equatorium in *Cosmographia ... Petri Apiani & Gemmae Frisii* (Antwerp: 1584).

the revival of learning in the Latin West. It is notable that a large part of the work is taken up with directions for making and using equatoria. Thus the equatorium entered western European astronomy right along with Ptolemy's planetary theory itself. [6]

The two circles are obviously eccentric to one another. The outer circle is divided into zodiac signs and is centered on point d, which represents the Earth. The inner circle is divided into equal 30° signs, starting from the apogee b of the eccentric circle. The center of the Sun's circle is point c, located slightly above d. To use the equatorium, one would first calculate the mean anomaly $\bar{\alpha}$, using tables to make the arithmetic easier. To obtain the Sun's longitude in the zodiac, one then pulls out a thread (no longer present) from the Earth d, stretches the thread through the computed position on the inner scale, and looks to see where the thread cuts the outer zodiac scale.

An even simpler equatorium is shown in Fig. 6. This comes from a sixteenth-century edition of the *Cosmographia* of Petrus Apianus, as modified by Gemma Frisius. This particular instrument was designed to look like the back of an astrolabe. This makes good sense, because the solar equatorium was a common feature on the backs of medieval European astrolabes. [7] In Fig. 6 the outer circle is divided into zodiac signs as usual. But on this instrument the eccentric inner circle is divided into months of the Julian calendar year.

There are two reasons why one cannot use today the equatorium of Fig. 6. First of all, it is calibrated in terms of the old (Julian) calendar, rather than the Gregorian calendar we use today. This is why the vernal equinox (zeroth degree of Aries) matches up with March 10—rather than with March 20, as it should today. Second, the longitude of the Sun's apogee changes slowly over the centuries. (The apogee moves forward in longitude at the rate of about 1.7° per century.) In the equatorium of Fig. 6, the apogee seems to be in the first few degrees of the sign of Cancer, which was approximately correct for the sixteenth century. Today, the apogee should be at about 103° of longitude, i.e., 13° into the sign of Cancer, as we saw above. [8]

Making a solar equatorium

It is easy to revise Apianus's solar equatorium so that it works for today's calendar and solar apogee, which is our third student project.

Make two photocopies of Fig. 6. We'll call them copy 1 and copy 2.

For copy 1: Cut out a circle of black paper of the right size to fill up everything inside the zodiac ring and paste this over copy 1. Thus, only the zodiac ring still shows; and the whole center of the diagram is black. Find the center of the disk and label it O (by noting the intersection of the line 0°Libra—0°Aries with the line 0°Capricorn—0°Cancer.) Use a pin to punch a hole through the center just found. Eventually you will thread a string through this hole from behind the disk. Draw a line from the hole in the direction 13°Cancer, to represent the direction of the Sun's apogee today.

For copy 2: Cut off the zodiac ring and the black ring just inside it and discard. Measure the radius of the Sun's circle. (Use the outer edge of the circle, with the alternating black and white 1°-segments). Multiply this radius by $e = 0.0334$, the eccentricity of the Sun's circle in the modern version of Hipparchos's theory. Let us call the product er. Also, find the center of the Sun's circle and label it C_2. Use a pin to punch a hole at C_2.

Back to copy 1: Measure out distance er from hole O along the line toward the apogee. Make a mark and label it C_1. Use a pin to punch a hole at C_1. Note that copy 1 now has two holes—one at O and one at C_1.

Putting the parts together: Stick a pin through copy 1 from behind, so that it pierces at C_1. Put copy 2 over the pin. The purpose of the pin is to guarantee that C_2 stays lined up with C_1 for the following step.

Turn copy 2 until July 5 lines up with the 13th degree of Cancer. (This is the Greenwich date—and place in the zodiac—when the sun reaches apogee.) Then glue copy 2 down to copy 1. When you are done, you can remove the pin.

Adding the fiducial string: Use a pin to push through the back of the hole at O, so that it goes all the way through both layers. Then thread a string through and knot it at the back so that it can't slip out. Be sure to leave five or six inches hanging free on the front side of the instrument.

You should now have an accurate solar equatorium for the twenty-first century. To use it to find the longitude of the Sun for any desired date, just pull the string through the date of interest and see where it cuts the zodiac. A good way to test the accuracy of your equatorium is to compare a few results with the answers (based on modern astronomical theory and computer calculation) that you find in in a current astronomical almanac.

Conclusion

The solar theory is an easily accessible branch of ancient planetary theory that can be mastered by students who have had some trigonometry. Even in the case of such a simple theory, the advantages offered by an equatorium are very striking. The equatorium allows one to visualize the theory in a most concrete way. And, it allows one to "calculate" the longitude of the Sun in the zodiac, simply by pulling a string through the day's date, rather than by undertaking a tedious trigonometric calculation. The equatorium also nicely bridges two common meanings of the word "model." The equatorium is a model made of paper and string. But it is also a concrete representation of the ancient Greeks' abstract theory (or "model") for the motion of the Sun. Both kinds of "modelling" were aspects of ancient Greek astronomy. The planets naturally pose more complications, with their epicycles for producing retrograde motion. But it is not too hard to build working equatoria for the planets as well. [9] The history of these instruments is also a richly rewarding topic of study. [10]

References

1. For the evolution of Greek planetary theory and detailed directions for using Ptolemy's form of it, see James Evans, *The History and Practice of Ancient Astronomy* (New York: Oxford University Press, 1998), Chapter 7.

2. Strabo (c. 64 B.C.–A.D. 25) says that students of geography should not be so ignorant as never to have seen a celestial globe or to have examined the circles drawn on it—tropics, equator and zodiac, etc. He says that this is the sort of knowledge one can acquire in the introductory astronomy courses. *The Geography of Strabo* vol. 1., trans. by H.L. Jones (London: Heinemann; Cambridge: Harvard University Press, 1969), i 1.21.

3. Geminos (first century B.C.) often refers to instruments and models (globes, sundials, and a sighting instrument called the *dioptra*) in the course of his elementary astronomy textbook. See James Evans and J. Lennart Berggren, *Geminos's* Introduction to the Phenomena: *A Translation and Study of a Hellenistic Survey of Astronomy* (Princeton: Princeton University Press, 2006), Introduction, Sec. 8.

4. [Nicolas] Halma, ed. and trans., *Commentaire de Théon d'Alexandrie sur les tables manuelles astronomiques de Ptolémée*, 3 vols. (Paris: 1822–1825), Vol. 1, p. 10.

5. C. Manitius, ed., *Procli Diadochi Hypotyposis astronomicarum positionum* (Leipzig: Teubner, 1909), iii, 66–72, p. 72–77.

6. Francis S. Benjamin, Jr. and G. J. Toomer, *Campanus of Novara and Medieval Planetary Theory* (Madison: University of Wisconsin Press, 1971).

7. For example, see Roderick and Marjorie Webster, *Western Astrolabes*, Historic Scientific Instruments of the Adler Planetarium & Astronomy Museum, vol. 1 (Chicago: Adler Planetarium & Astronomy Museum, 1998).

8. We obtained 102.4° for the longitude of the apogee. The season lengths that were given above, and on which this figure for A is based, were valid for the 1970s. The apogee continues to move forward at its stately rate of 1.7°/century. So, for the year 2005, a good value is $A = 102.9°$, or 103° to the nearest whole degree.

9. For directions for making and using an equatorium for Mars, see James Evans, *The History and Practice of Ancient Astronomy* (New York: Oxford University Press, 1998), p. 406–410.

10. For an introduction to the history of the equatorium see John North, *Richard of Wallingford* (Oxford: Oxford University Press, 1976). A comprehensive study of the later period is provided by Emmanuel Poulle, *Les instruments de la théorie des planètes selon Ptolémée. Equatories et horlogerie planétaire du XIIIe au XVIe siècle* (Geneva and Paris: Librairie Droz, 1980). For the equatorium attributed to Chaucer see Derek J. De Solla Price, *The Equatorie of the Planetis* (Cambridge: Cambridge University Press, 1955). For the sumptuous Renaissance equatoria of Petrus Apianus see Owen Gingerich, "Apianus's Astronomicum Caesareum and its Leipzig Facsimile," Journal for the History of Astronomy 2 (1971) 168–177 and S. A. Ionides, "Caesar's Astronomy (Astronomicum Caesareum) by Peter Apian, Ingolstadt 1540," Osiris **1** (1936) 356–389.

Sundials: An Introduction to Their History, Design, and Construction

J. L. Berggren[1]
Simon Fraser University

"If you put your nose pointing to the sun and open your mouth wide you will show every passerby the time of day." Quoted from *The Greek Anthology* in [2. p. 3]

Introduction

The sundial is one of mankind's oldest instruments for telling the time during daylight hours. The earliest surviving dials come from Egypt. There we find dials from as early as the 15th century B.C. with a short vertical block (called a gnomon)[2] of finished stone at the end of a horizontal stone ruler marked with a scale of hours [6, p. 59]. When the device is turned so the gnomon faces the sun and casts its shadow on the ruler, the end of the shadow shows the hour of the day according to an approximate arithmetic scheme.

The first people to find exact methods for using shadows to tell time were the ancient Greeks, who used geometry and a geometrical model of the cosmos to construct sundials.[3] Aristarchus of Samos[4] invented a sundial called "the bowl" or "hemisphere," which consisted of a block of stone with a hemisphere hollowed out of the top. From the top rim of the hemisphere a thin rod extended to the center. When the sun was shining (as it often does in Greece!) the tip of the rod's shadow moved across a net of lines on the hemisphere below and told the time as the sun moved across the hemisphere of the sky. Berosus the Chaldean modified this to form a dial by cutting away the part of the hemisphere south of the course of the shadow tip at the summer solstice.[5] (Figures 1a and 1b show the original bowl dial and Berosus's cut-away version.)

The hemispherical dial vividly illustrates the point that the design of a sundial rests on a mathematical projection of the surface of the visible hemisphere onto some other surface. In these cases it is a hemisphere, but Greeks also designed dials in which hour lines appear on the surfaces of cones, cylinders, globes, and planes[6].

---

[1] The author thanks Dover Publications for generously allowing the reproduction of figures from [3]. He also thanks Brian Albinson, Margareta Hedin, Siegfried Rasper, Fred W. Sawyer, III, and the editor, Amy Shell-Gellasch, for their contributions to the writing of this paper. Any errors that remain are the author's responsibility.

[2] A Greek word denoting, in this context, a vertical object, such as a block or rod, erected specifically to cast a shadow.

[3] For a good account of Greek and Roman sundials see [2].

[4] A Greek astronomer, best known for exploring the hypothesis of a heliocentric cosmos. Ptolemy credits him in his *Almagest* with observing the summer solstice in 281/80 B.C.

[5] Berosus worked on the Aegean island of Cos ca. 270 B.C. That he and Aristarchus were contemporaries, lends credence to these two reports of early Greek sundials. That Berosus's dial was a modification of Aristarchus's dial, and not an independent invention, is simply my supposition. (Book IX of Vitruvius's *Architecture*, is our source for this, and he simply lists both men as inventors of dials.)

[6] For pictures of some known conical, globe and planar dials see [2, 231 (conical); 376–378 (globe); 330 (planar)]. [2, 56] also refers to five cylindrical dials of Roman provenance.

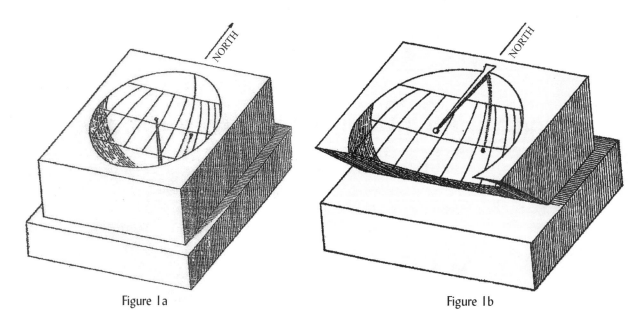

Figure 1a   Figure 1b

Greek dials differed from modern dials in three ways: (1) They told time using the concept of a "seasonal hour", a period of time equal to 1/12 the length of daylight on a given day.[7] (2) They commonly had horizontal or vertical gnomons, rather than one pointing to the north celestial pole, as is usual today. (3) They indicated time by the position of the **tip** of the shadow of the gnomon, not the **edge** of its shadow.

The Greek theory of the cosmos

Since the ancient Greek theory of the cosmos is still a useful pedagogic device today for helping students understand how a sundial works we shall explain it briefly here.[8]

Greek astronomers believed the universe, which they called the 'cosmos,' was an immense, but finite, sphere[9], which rotated once each day around an axis through the north and south celestial poles and the center of an immobile earth (shown in gray in Figure 2), whose center coincided with that of the cosmos. (We shall, for brevity, call the surface of the sphere of the cosmos the 'celestial sphere.')

The earth was also a sphere, but so small in relation to that of the cosmos that it had the ratio of a point to the size of the larger sphere.[10] A consequence of this is what we call the:

Fundamental Theorem of Gnomonics. *When designing a sundial we may take any point on earth to be the center of the cosmos.*

The cosmos rotated each day in a westerly direction uniformly around an axis that joined the north and south celestial poles[11]. (We shall henceforth refer to it as 'the polar axis.') As the cosmos rotated, it carried the sun, moon, and stars around the earth. And the sun was so far away that all its rays hitting the earth at any moment could be considered as being parallel.

In this model three great circles are important, all of them imagined as being on the celestial sphere. (For the Greeks, circles were two-dimensional areas and the intersection of these circles with the earth at the center of the cosmos gave rise to like-named circles on the earth.)

---

[7] Their lengths depend not only on the local latitude but also on the season of the year, hence the term 'seasonal'.

[8] For a discussion of the cosmography relevant to dials from a heliocentric standpoint see [5, pp. 18–31].

[9] The stars were in the surface of the sphere and were carried around by the daily rotation of the cosmos.

[10] See, for example, Ptolemy's *Almagest* I,6.

[11] Today the North Star is near the northern end of this axis. But, in Euclid's time for example, there was no visible star near this spot.

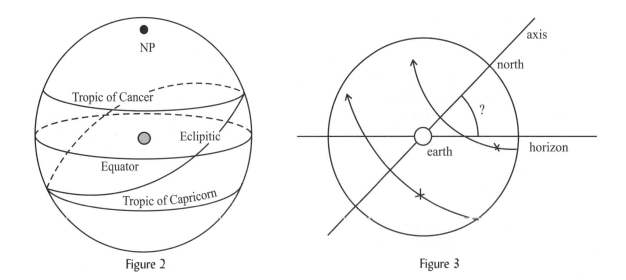

Figure 2   Figure 3

One of these circles is the equator (shown in Figure 2), a great circle on the celestial sphere. Perpendicular to the polar axis, it separates the northern hemisphere from the southern.

The horizon (seen in section as a straight line in Figure 3) of any locality on the earth is the intersection of the celestial sphere with the plane tangent to the sphere of the earth at that locality.

The horizon separates the part of the cosmos visible at that locality from the invisible part. By the Fundamental Theorem of Gnomonics we may assume that the horizon plane passes through the center of the cosmos and so, the horizon is a great circle on the celestial sphere.[12]

This circle is a fixed plane, inclined to the equator at an angle equal to the complement of the observer's latitude. Hence, as Figure 3 shows, the angle between the polar axis and the horizon is equal to the latitude of the locality (denoted by ϕ), a basic fact for dial makers.

A third great circle relevant to the design of sundials is the local meridian, which passes through both poles and the point directly above the observer, called the zenith.[13] (The intersection of the meridian with the earth is what is nowadays also called, in geographical parlance, the local meridian.) The meridian circle is perpendicular to the horizon and equator, and its intersection with the earth at a locality marks the local (true) north and south. The sun reaches its maximum altitude[14] for any given day, and is due south for an observer north of the Tropic of Cancer, as it crosses the local meridian.

In addition to these great circles, other circles are also important. In the course of a year the sun appears to make a rotation around the celestial sphere, and its annual path as seen against the background of that sphere is a circle on that sphere, called 'the zodiac' or 'ecliptic'. This is shown in Figure 2 as a circle skewed to the equator and bounded by two circles parallel to the equator. One, on the north, is known as the Tropic of Cancer and that on the south is called the Tropic of Capricorn. On any given day the sun is at some position in one of the twelve signs of the zodiac, and in the course of the daily revolution of the cosmos around the earth the sun makes, to a first approximation, a circle centered on the polar axis and perpendicular to it.[15] This is called its *day circle*. These circles are parallel to the equator and are sometimes referred to as

---

[12] We may, thus, speak of the visible and the invisible *hemispheres* of the cosmos for a given locality.

[13] The great circle on the earth directly below it in this model is the meridian of longitude of the observer.

[14] The altitude of any celestial object (relative to a horizon) is the arc of the great circle passing through the zenith of the observer between the object and the horizon.

[15] This is true only to a first approximation since, as the ancients would have put it, the sun also has a slow *eastwards* motion of its own along the circle of the zodiac amounting to a little less than 1° per day. Since the zodiac is not perpendicular to the polar axis the motion of the sun on the zodiac means that the sun appears to move north and south in the sky during the course of a year, so that in fact the sun during the day traces out part of a spherical helix.

parallel circles. Only one of them, the equator itself, is a great circle. Figure 3 shows arcs of the day circles of the sun at two different positions, denoted by small crosses.

Following upon the Greek invention of the science of sundials (called 'gnomonics') both the medieval Islamic civilizations and Renaissance Europe made important contributions to its development. Islamic astronomers, in particular, invented dials whose gnomons pointed to the north celestial pole.[16] This idea, which connected the design of dials with the apparent motion of the sun around the cosmos during the course of the day, had important consequences for the design of dials, which we shall now elaborate.

The equatorial dial

The geometry of the celestial sphere with its axis, equator, and meridians, enters the design of a sundial for a given locality as follows. At midnight (sun time) the sun will by definition be on the northern half of the meridian of the locality. At noon it will have traveled halfway around its day circle so it will be on the southern half of that meridian. Midway between these two times, at 6:00 A.M., the sun will have traveled one-fourth of its day circle, and so it will be east of the locality. Similarly, at 6 P.M. it will be west of the locality, three-quarters of the way around its day circle. By midnight it will be back where it started.[17]

With this in mind it is easy to understand the design of our first dial, called an *equatorial dial*, two examples of which are shown in Figure 4, from [3, pp. 22 & 55].[18] (Note, however, that the direction of celestial north is up and 'into' the paper in the dial on the left, but up and 'out of' the paper in the dial on the right.)

Equatorial dials are most aptly described as "bare-bones" models of the cosmos relative to the user's locality. The dial on the left consists of a vertical semicircle, representing the northern half of the observer's meridian. To it is fastened, at right angles, a graduated semicircle, which represents the plane of the equator. (Both semicircles may be supposed to be in the planes of the circles they represent.) The equator has a scale of hours around it, usually from 6:00 A.M. on the west edge to 6:00 P.M. on the east, and these may be subdivided into 5- or 10-minute intervals. A gnomon joins the north and south poles on the meridian circle, and thus represents the polar axis. It is tilted to the horizon at an angle equal to the latitude, so it is parallel to the polar axis (Again, see Figure 3). So, by the Fundamental Theorem of Gnomonics, the intersection of the diameters of the two perpendicular semicircles of the dial may be taken as the center of the cosmos, and the gnomon as lying in the polar axis.

Figure 4

---

[16] Space does not allow a proper account of the many contributions of Islamic astronomers to sundials, and the reader who wants to read something of them should consult [1].

[17] This neglects the slight daily motion of the sun north or south of the ecliptic discussed in Note 15.

[18] A large example of such a dial may be seen at http://web.fcnet.fr/frb/sundials/photos/chicago.jpg.

As the sun travels across the sky from east to west each day, the shadow of the gnomon tracks steadily across the equatorial semicircle on the dial, from west to east.[19] And at any time during the day it has tracked across the same proportion of the equatorial semicircle as the sun in the same time has moved along its day circle across the sky, but in the opposite direction.

This dial is based on an image of the celestial sphere divided into 24 equal segments, like an orange, by 12 great circles through the poles and forming equally spaced meridians. The equator plane is perpendicular to the polar axis (which is the common intersection of these great circles) and the horizon plane is skewed to it.

Making your own sundial

To make your own equatorial sundial, in which the shadow falls on a circular disk parallel to the equator as on the right in Figure 4, photocopy Figure 5 onto a piece of heavy-duty construction paper, and assemble the kit as shown in Figures 6a and 6b.

Make the indicated cuts (solid lines) and folds (dashed lines).[20] Fold the two sides of the gnomon and glue them onto each other. In Figures 6a and 6b the pointed end of the gnomon sticks upwards through a slit in the face of the dial, the gnomon being perpendicular to the face.

The lower edge of the gnomon passes through the dial face so it just touches the line joining the marks for 6 A.M. and 6 P.M. As a consequence it is the lower edge of the shadow that tells the time.

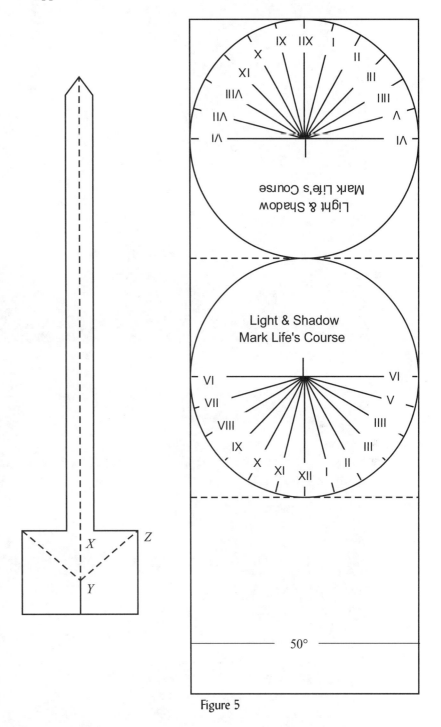

Figure 5

---

[19] For this reason the morning hours appear on the western edge of the equatorial arc.

[20] One does not, however, cut around the solid circumference of the circle or any of the hour lines inside it. The solid lines next to the '50' mark the positions of the lower edge of the gnomon for that latitude. (The dial in the photo has numerals marking corresponding positions for other latitudes.)

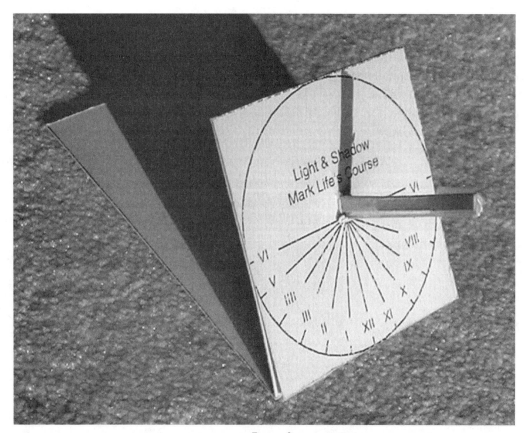

Figure 6a

Figure 6b

The dial is designed for 50° N. latitude but can easily be adapted to any latitude by making angle XYZ in Fig. 5 equal to the latitude of the locality. The dial has numerals on both its upper and lower faces because the dial faces are in the plane of the equator.[21] So when the sun is north of the equator, in spring and summer, from (roughly) March 21 to Sept 21 the shadow of the gnomon falls on the upper face, but during the rest of the year it falls on the lower face.

Sundials and clock time

But can such a simple device really do what our clocks do, i.e. tell accurate time? The answer to this question depends on what kind of time we mean.[22] Our equatorial dial shows true local sun time. For example, it will show noon when the sun is due south. But most people will find that the sun is not due south when their watches show noon. This has nothing to do with whether they are using daylight saving time but arises from the difference between sun time and standard time.

Until the late 1800s, standard time had not been invented and time at each locality was governed by the sun's position relative to the local meridian. A good sundial was more accurate than a good clock, and up to 1900 the French railroad system used precision sundials, rather like the example on the left in Figure 4, known as heliochronometers, to set its clocks [5, p. 17]!

However, the introduction of the railroad, telegraph, and other means of rapid communication made local time an inconvenience. And it was a railway planner, the Canadian Sir Sandford Fleming, who outlined a plan for worldwide standard time in the late 1870s.[23] Following Fleming's ideas, delegates from 27 nations met in 1884 in Washington, D.C., for the Meridian Conference and agreed on a system of standard time that is basically the one now in use.

Twenty-four standard meridians of longitude, 15° apart, form the framework of standard time, the first meridian (through Greenwich, England, at longitude 0°) being the prime meridian. These meridians are the center lines of 24 standard time zones, and each is referred to as the meridian for its zone. For example, Vancouver, B.C., whose longitude is 123°, is in the Pacific Standard Time zone, whose meridian is 120° W of Greenwich.[24]

Since each degree of longitude corresponds to 4 minutes difference of time (24*60/360) one can correct the time read from an equatorial dial to obtain standard time by adding (subtracting) 4 minutes from the time shown on the dial for each degree of longitude one is west (east) of the meridian for one's zone. One could, equally well, rotate the hour markings on the equatorial band around the gnomon counterclockwise (clockwise) an appropriate number of degrees and so build the correction into the dial.

However, as we indicated above, our model is only a first approximation, because the westward motion of the sun during the course of a day is not exactly uniform. As we said, the sun also moves slowly eastward along the ecliptic[25] (shown in Figure 2) over the period of a year. This motion varies but is, on the average, slightly less than 1° per day. In addition to this, the ecliptic is inclined to the equator. These two factors combine to create a disagreement between a sundial and watch time, which varies according to the time of year, but may amount to more than 16 minutes at some times. For example, on Halloween a sundial is more

---

[21] This follows from the Fundamental Theorem of Gnomonics and the fact that the gnomon makes an angle with the horizon equal to the latitude.

[22] For another discussion of the different kinds of time see [7, pp. 6 – 17].

[23] For details on Fleming and standard time see http://geography.about.com/library/weekly/aa030899.htm.

[24] In practice, the theoretical zone lines have in many cases been adjusted for the convenience of inhabitants. Thus all of China, which stretches from E to W over 4 theoretical time zones, is on the same time. And Spain, most of which is west of the Greenwich meridian, is on Central European time, where the clocks are one hour ahead of Greenwich.

[25] This is a great circle running along the middle of the belt of constellations known as the zodiac.

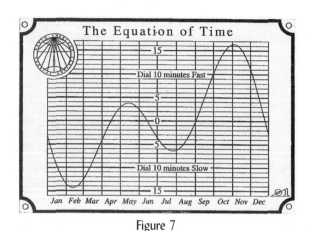

Figure 7

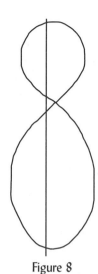

Figure 8

than 15 minutes ahead of a watch. This difference is called 'the equation of time' and Figure 7 shows how it varies over the year.[26]

As with the correction for standard time, the equation of time may be incorporated into the design of a dial by means of a curve called the analemma,[27] an elongated, asymmetric "figure-8." shown below in Figure 8. This curve is divided into 12 segments (not shown here), representing the 12 months of the year. The curve has a slightly off-center vertical axis that passes through points representing the four days during the year (as shown in Figure 8) when the equation of time is zero.

The perpendicular distance to the axis from a point in a segment corresponding to a given day of a given month measures, in an appropriate unit, the magnitude of the equation of time.[28] According to the French historian of astronomy, J. B. Delambre, the French gnomonist Grandjean de Fouchy (1740) first constructed this curve. (It, or the graph, is often seen on sundials.) Figure 9 shows an equatorial sundial built by the talented gnomonist Brian Albinson of North Vancouver. It has not only a graph of the equation of time on the white vertical piece in back but also has the characteristic "figure 8" of the analemma cut into the gnomon in such a way that when the gnomon is facing the sun the dial automatically corrects for the equation of time. The gnomon can also be rotated 90°, as in the photo, to show sun time, indicated by the center of the sunbeam passing through a slit in the gnomon.

Figure 9

---

[26] This plate, published by the North American Sundial Society, shows that the sundial is furthest behind the clock in mid-February and furthest ahead at the beginning of November.

[27] The vertical axis is measured in 5-minute intervals from –15 minutes to +20.

[28] The use of the word 'equation' here is traditional but reflects an archaic usage of that term to denote a correction added to or subtracted from a first approximation to produce a closer approximation.

Failure to understand the corrections for mean time and the equation of time leads many to conclude with Hilaire Belloc that a good motto for sundials is: "I am a sundial and I make a botch/Of what is done much better by a watch."[29] However, a well-constructed dial tells local sun time very accurately, and, with the corrections just discussed, will yield time as shown on a watch as well.

Other Dials

With an understanding of the equatorial dial it is easy to see how to design the familiar **horizontal dial**, shown in Figure 10 [3, p. 25]. In this dial, the gnomon is the top edge of what is usually a triangular-shaped piece mounted perpendicularly on the horizontal base of the dial.

Figure 10

One edge of the gnomon (the one culminating in the upper point in Figure 10) makes an angle ϕ with the base, and so points towards the celestial north pole. On the dial are inscribed the numerals for the hours and the edge of the shadow falling on the hour lines indicates the time of day. Unlike successive hour lines on the equatorial dial, those on the horizontal dial do not make equal angles. Rather, the angles between the successive hour lines depend on the local latitude, i.e. the angle the gnomon makes with the base. This is because the horizon is not perpendicular to the earth's axis but is skewed to it. And finding these angles is the principle mathematical task in designing a horizontal dial.

In making a horizontal dial one may have to take into account the thickness of the gnomon. This is because it is the western edge of the gnomon that casts the morning shadows and the eastern edge that casts the afternoon shadows. So, if the gnomon has any substantial thickness, the dial should have two noon lines, parallel to each other and separated by the width of the gnomon. (In Figure 10 the sloping edge of the

---

[29] Quoted from the article "Can your sundial really tell the time?" by Hugh Harris-Evans on the website www.goarticles.com/cgi-bin/showa.cgi?C=24342.

gnomon which casts the shadow is beveled to form a line, so the same edge casts the shadow all day and there is no need for two noon lines. But, often, the horizontal dial is a split dial, with a morning half and an afternoon half.)

The formula for the angles, A, that the hour lines on a horizontal dial make with the noon lines is $\tan(A) = \tan(t)\sin(\phi)$, where t is the so-called "hour-angle" (15° for 11 A.M. or 1 P.M., 30° for 10 or 2, etc.), and ϕ is the local latitude. Thus, for Vancouver, B.C., at latitude 49° north, the hour line for 10 A.M. should be drawn at an angle of 23.5° to the noon line.

The two main mathematical tools in dial making are calculation and geometrical construction. A teacher electing to approach the subject through the former can give the student interesting, and often non-trivial, applications of trigonometry, and in this electronic age, calculation is a quicker approach. Some geometric constructions, however, especially when explained in terms of the intuitive model of the equatorial dial, can be more illuminating for the student, and their derivation can develop a student's spatial intuition. The exercises at the end of this paper suggest possibilities for both approaches.

In [3, 99-1-1] and [4,45-47] there are instructions for constructing the hour lines for a horizontal dial of given latitude by plane geometry, without doing any calculation. An outline of the construction, based on the description in [4], is as follows. One first draws three parallel lines, JJ', NN' and MM'. In doing this construction one chooses the distance, LN, between the lower parallels to be any convenient length for the base of the gnomon. Then one takes KN, the distance between the upper parallels, to be $LN*\sin(\phi)$. Finally, one constructs KL and $K'L'$ perpendicular to JJ' and separated by a width that one wants to use for the gnomon of the sundial.

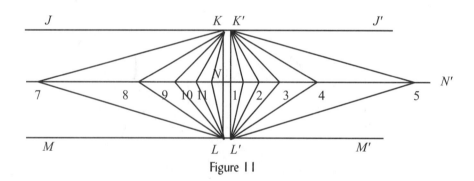

Figure 11

Having thus established the basic framework for designing the dial, one now draws lines radiating from K and K' so they divide each of the two right angles at K and K' into 6 equal parts of 15° each. Then one uses the numerals shown in Figure 12 to label the intersections of these lines with line NN'.[30] Finally, one joins these intersections with the points L and L' as shown in the figure. These lines are now the hour lines for a horizontal sundial for a locality having latitude ϕ and whose gnomon, of width LL', points toward the pole in the direction of N.

To see why this is so, imagine the part of Figure 11 above the line NN' to be creased along that line and folded towards the reader until the two strips of paper make an angle of $90° - \phi$ at the crease. It is a good exercise to show that if L and K are joined by a line, as in Figure 12, then angle L would equal ϕ and angle K would be right. Thus the strip KN, which has the hour lines for an equatorial dial radiating out of K, would be set up exactly right for an equatorial dial with the (imaginary) piece KL as the part

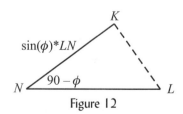

Figure 12

---

[30] Note that the noon lines, the two parallels through L and L', are not labeled "12" here, for lack of space.

of its gnomon between the dial and the horizon.[31] (See the dial on the right in Figure 4.) The reader who wants a challenge may now complete the argument to show that the lines on the lower dial are indeed the hour lines for a horizontal dial for latitude ϕ and gnomon KL.

Just as we can design a horizontal dial on the basis of an equatorial dial, so we can design a **vertical dial** on the basis of a horizontal dial. ("Vertical dial" refers to one mounted on a vertical wall, here assumed to be facing south. Its gnomon is parallel to the polar axis.) Since the earth is a sphere, a vertical dial at latitude ϕ North will, if moved parallel to itself southward through an arc of 90° along the same meridian, be a horizontal dial at latitude $90° - \phi$ South. (See Figure 13, where Z denotes the zenith of a locality in the northern hemisphere.)

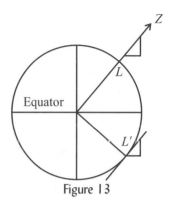

Figure 13

By the Fundamental Theorem of Gnomonics the dial shows the same time at either place. And, because we moved the dial along a meridian, the time *is* the same at either place. Hence, to design a vertical dial for latitude ϕ North one need only design a horizontal dial for latitude $90° - \phi$ South. (Note that the gnomon of such a dial points to the south celestial pole.)

Sundials as calendars

The ancient Greek gnomonists realized that sundials could be used to indicate not only the time of day but also the sign of the zodiac in which the sun is located on a given date. In Figure 1a notice that three circular arcs run across the hour-lines, arcs which indicate the seasons of the year. The northernmost of the three arcs is the path of the tip of the gnomon's shadow when the sun enters the sign of Capricorn (i.e., the winter solstice). The southernmost arc is the path of the tip of the gnomon's shadow when the sun enters Cancer (i.e., the summer solstice). And the middle arc is the path when the sun enters the signs of Aries or Libra (the vernal and autumnal equinoxes).

In fact, anything dependent only on the declination of the sun above or below the equator, such as the beginnings of the months or special dates, such as anniversaries or birthdays, could be indicated on a dial. Thus, on the vertical dial shown in Figure 14, from St. Mary's Church in Krakow, Poland, the shadow of the tip of the gnomon indicates the beginning of the months as it traces out the various hyperbolas.[32]

Figure 14

---

[31] Note that the width of the gnomon means this equatorial dial is also 'split,' so that, e.g., the western edge of the gnomon casts the shadow for morning hours.

[32] Unlike the circular lines in Figure 1 the curves representing the months in Fig. 12 are hyperbolas. This is because the gnomon tip is projecting part of the sun's day circle onto a plane surface at an acute angle to that circle. (I thank my friend Siegfried Rasper, of Nienhagen, Germany, for this photograph.)

Sundials in the classroom

In giving talks to high school and university students on the topic of sundials I have found that students are fascinated by the question of how people told time before the invention of the clock. Most of them have heard about sundials and know they are connected with astronomy, but they are invariably surprised to learn that the design of sundials relies heavily on geometry and trigonometry, so sundials are also part of the history of applied mathematics. Because a range of mathematics problems—from the simple to the difficult—arise in the design and use of sundials they are a good source of problems for classes of varying abilities.

To give an idea of the range of problems that a teacher can use we state a few here.

Problem 1. A family traveling comes on a charming little town whose central square has a large bubbling fountain in the center as well as a beautiful sundial. The children recognize the dial as an equatorial dial but are surprised that its gnomon is parallel to the surface of the square. Suddenly the big sister grabs her little brother and dunks him in the fountain. Why did she do that?

Problem 2. On what days of the year will a dial at your school such as that shown in Figure 11 show the same time as an accurate watch does? [Hint: To solve this the student needs to use the analemma curve shown in Figure 8 and must know both the local longitude and that of the meridian for his/her time zone.]

Problem 3. At what days of the year, and hours of the day, would a vertical dial for the north face of a house be useful in telling time? Make a sketch showing how the appearance of such a dial would differ from that of a dial for the south face of the same house.

Problem 4. A student makes a horizontal dial whose gnomon is simply a stiff bar of metal or a piece of small-diameter pipe inclined at the appropriate angle. Thus the shadow it casts on the dial always has two sides. Which side of the shadow should the student use at 8 A.M.? What about noon? What about at 3:00 P.M.?

Problem 5. Using trigonometry, design the hour lines of a horizontal dial for your school with a gnomon one centimeter wide. Design a vertical south-facing dial as well.

Problem 6. Use the elementary theory of conic sections and a rough sketch to show that in general the tip of the shadow on a horizontal dial, during any given day, traces out a hyperbola. [Hint: Think how the parallel circle that the sun travels on during the day and the tip of the gnomon give rise to a cone that will be cut by the horizon plane.] Discuss possible exceptions.

Problem 7. Use the references below by Rohr and Waugh, as well as the web resources, to write a report about analemmatic dials. It should explain the most obvious difference between such dials and those we have discussed in this article, tell when and where they were probably invented, and why the numerals for the hours on such dials are arranged on the circumference of an ellipse. Now use the sundial software provided by websites [10] or [12] to design such a dial for your school.

Conclusion

I hope that this paper will stimulate teachers to include some discussion of the basic geometry of simple sundials in their geometry and trigonometry classes. The theory uses significant high school level material from both these subjects, but not so advanced as to be beyond the reach of most students. It also gives the mathematics teacher a chance to talk about the geometry of the surface of the sphere, which provides a nice introduction to non-Euclidean geometry (something, I have discovered, that interests both teachers and students). Since the topic has a rich history it ties in nicely with social studies and art, and the construction of a sundial for the school ground would make an excellent class project in which both mathematics and art classes could leave a heritage for coming generations of students.[33]

References

1. Berggren, J. L. "Sundials in Medieval Islamic Science & Civilization." *The Compendium*. Vol. 8, No. 2, 2001.
2. Gibbs, Sharon L. *Greek and Roman Sundials*. New Haven, Conn.: Yale U. Press, 1976.
3. Mayall, R. Newton and Margaret L. *Sundials: How to Know, Use, and Make Them.* Boston: Hale, Cushman and Flint, 1938. (Reprinted by Dover as *Sundials*: *Their Construction and Use*.)
4. Milham, Willis I. *Time and Timekeepers*. New York: The Macmillan Company, 1947.
5. Rohr, René R.J. *Sundials: History, Theory, and Practice*. Dover, 1996.
6. Turner, Anthony J. (ed.), *Time* Amsterdam, 1990.
7. Waugh, Albert E. *Sundials: Their Theory and Construction*. Dover, 1973.
8. http://www.sundials.co.uk/home3.htm Besides having much useful information, and names of commercial outlets where you can buy sundials or even have one custom made for you, this is also a useful gateway to the many international sites featuring sundials.
9. http://www.sundials.org This is the official website of the North American Sundial Society. For quite modest dues you can join, receive their journal, *The Compendium*, and participate in their annual meetings.
10. http://home.iae.nl/users/ferdv/index-fer.htm This is the site of the Dutch gnomonist, Fer De Vries, from which you can download free Windows software to design your own sundials.
11. http://members.aon.at/sundials/index-e.htm This site shows a beautiful collection of color photos and descriptions of sundials from many lands, along with other material on dials, by the Austrian gnomonist Karl Schwartzinger.
12. http://web.utanet.at/sondereh/sun.htm This is the site of the Austrian gnomonist, Helmut Sonderegger, from which you can download free Windows software to design your own sundials.

---

[33] The North American Sundial Society has a public outreach program which, in some areas of the country, can provide an experienced dialist to help in the planning and design of a sundial for your school. The Society can be reached through its website, whose url is listed in #9 above.

Why is a Square Square and a Cube Cubical?

Amy Shell-Gellasch
Pacific Lutheran University

Introduction

Why is the algebraic process of forming a perfect quadratic expression referred to as completing the square? With simple cutouts for an overhead projector, the geometric underpinnings as well as many aspects of quadratic equations can be exhibited quickly and effectively. This can be expanded to the cubic with the use of five wooden blocks. This chapter will show how to use these items in the classroom to give students a geometrically intuitive as well as an historical understanding of quadratic and cubic expressions.

Creating squares

At several points in the undergraduate curriculum, as well as in some secondary courses, we find ourselves "completing the square." I found myself in just such a position a short while back while teaching a sophomore level calculus course. On the spur of the moment, I asked the class if they knew why it was called "completing the square". The response ranged from blank stares to shrugged shoulders. Looking at the clock, I decided I could sidetrack for a few minutes.

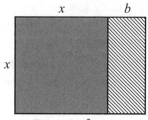
Figure 1. $x^2 + bx$

I started by drawing the Figure 1 on the board to represent the statement $x^2 + bx$. (In general, you will have some expression of the form $ax^2 + bx + c$. But the actual process of completing the square is done on an expression of the form $x^2 + bx$, after division by a and "removal" of c.) I start with the comment that for the ancient Greeks as well as the Babylonians, geometry was physical; that x (in modern notation) represented a length, x^2 an area, and so on.

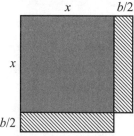
Figure 2. $x^2 + 2(b/2)x$

The next step in completing the square requires $b/2$ to be calculated. (I draw as the students recite the steps in the process of completing the square.) By cutting and pasting in the Chinese style[1], you have the same area, but in the form $x^2 + 2(b/2)x$, as in Figure 2.

Finally you add $(b/2)^2$, represented by the small square at the bottom right of Figure 3.

You now have regions with areas:

$$x^2 + bx + (b/2)^2 = (x + b/2)^2.$$

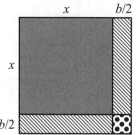
Figure 3. $x^2 + 2(b/2)x + (b/2)^2$

---
[1] See David Zitarelli's chapter on the Chinese method for more on this point.

So you have "completed the square" quite literally. The above equation and diagram allows students to *see* why we need the "square of half the second term" when completing the square. By having your students solve for *x* in this equation, they will understand where the quadratic equation comes from.

This can be done quickly at the board with chalk in one hand and eraser in the other. However, precut and colored pieces of overhead transparencies can be used to much more effect. This can also be done anytime the topic comes up, but it is most effective as a transition to a discussion of the cubic.

I have found that most students can recite the steps in the process most successfully, but have no idea where they come from. This quick demonstration is all that is usually needed to make the steps as well as the name clear. From my experience, this quick side-trip into the history of mathematics is very enlightening and exciting for many students. Since it can be done so quickly, and on a topic that they are familiar with, it is a painless way to incorporate some history into your teaching. (Inserting history of mathematics into a lesson just for the sake of incorporating it can seem arbitrary to the students, and as such, it becomes a burden to the teacher and the students if overdone. However, inserting history at those "teachable moments" whether planned or not, is more natural and profitable for the students.)

A little history

The ancient Babylonians describe a way of "completing the square" [2] that is a cut and paste method that transforms a rectangle to a square. [2] This justification for the "algebraic" method of completing the square resurfaces two and a half millennia later with the Islamic scholars. The Babylonian process is almost identical to that provided by Al-Khwarizmi in the 9th century. Arab mathematicians, like their Greek forerunners, did not acknowledge negative coefficients or roots. Each quadratic was categorized and solved according to its form. For example, $ax^2 = bx + c$ and $ax^2 + bx = c$ (with a, b, c all positive) were considered different forms. Al-Khwarizmi was dealing only with the form $x^2 + bx = c$. In particular the case $x^2 + 10x = 39$.[3] [1] As was the norm, he stated his solution in narrative form. It translates algebraically to

$$x = \sqrt{(b/2)^2 + c} - b/c.$$

To explain his solution, Al-Khwarizmi describes a physical cut-and-paste method equivalent to that presented above. Using a geometric or physical interpretation to justify or explain a non-physical or algebraic process was very rare at that time. Thus Al-Khwarizmi was ahead of his time by using tools from geometry to solve algebraic problems. It was not until the 17th century that solving the general quadratic in the form $ax^2 + bx + c = 0$ was done without dealing with individual cases of the coefficients.[4]

To finish the story of my classroom experience: at this point, one of my students, who had been an average, though uninspired and for the most part uninvolved student, really perked up. That the term "completing the square" literally meant something intrigued him. When I then explained that to the ancient Greeks virtually everything in geometry had a physical meaning, he was hooked. Not only did his grades and involvement improve, his new interest in math history motivated him to submit a paper for the Association for Women in Mathematics writing contest that won him first prize at the collegiate level!

---

[2] Remember that the term completing the square is a modern construct; and the Babylonians did their mathematics rhetorically. Symbolic algebra is only approximately 500 years old.

[3] Babylonian mathematics, as well as that of many other ancient cultures, did not present general procedures as we are accustomed to. They presented a specific, though general example to show the method. Students were expected to glean the general procedure from one or more of these examples.

[4] For general information on this topic see a general history of mathematics text such as Victor Katz's *A History of Mathematics: An Introduction*, for more detail see Ronald Calinger's *Classics of Mathematics* [3].

The cubic

A brief history

Virtually any history of mathematics book from popular accounts such as [4] to scholarly texts such as [2] visits the very public and controversial dispute between Cardano and Tartaglia over the solution to the general cubic (c. 1545). So how did Cardano solve the cubic? And does it have a physical manifestation as the solution to the quadratic does? The method presented here is a modernized version of what Cardano did, and is equivalent to Al-Khwarizmi's work. See Ronald Calinger's *Classics of Mathematics* for their original works. [3] As stated above, since negative numbers in general, and negative roots to equations in particular, were not accepted until the Renaissance, mathematicians from the ancients through Cardano considered separate cases where we would consider only one. We would write all cubic equations as $ax^3 + bx^2 + cx + d = 0$, with all coefficients ranging over the reals. Mathematicians such as Cardano wrote the equations so that all coefficients were positive. In particular he solved the cases of *a cube and a line equal to a number* and *a cube equal to a line and a number*. In modern algebraic notation this would be $x^3 + cx = d$ and $x^3 = cx + d$ respectively.[5]

Solving the cube

So now, how would you solve or complete the cube? First, since the solution to the quadratic was already known from antiquity, Gerolamo Cardano (1501–1576) wanted to transform the cubic into a quadratic, or at least a form susceptible to the quadratic formula. He realized that an equation of the form $x^3 + px = q$ can be rewritten in the form $u^6 + ru^3 + s^3 = 0$, which is a quadratic in u^3. His line of attack was then to transform the general cubic into such a quadratic.

Consider a cube with side dimension $a + b$ (see Figure 4).

$$(a+b)^3 = a^3 + 3a^2b + 3ab^2 + b^3 \tag{1}$$

Regroup to get

$$(a+b)^3 = a^3 + 3ab(a+b) + b^3. \tag{2}$$

Now let $a + b = u$ and $b = v$. Then $a = u - v$ as in Figure 4.
Rewriting (2) produces

$$u^3 = (u-v)^3 + 3uv(u-v) + v^3. \tag{3}$$

Cardano used the following clever substitutions:

$$x = u - v, \tag{a}$$

$$p = 3uv, \tag{b}$$

$$q = u^3 - v^3. \tag{c}$$

After rearranging terms, Cardano transformed (3) into $x^3 + px = q$, where p and q are positive. Using equation (b) which defines p, we have

$$v = \frac{p}{3u}.$$

From (c), this leads to

$$q = u^3 - \left(\frac{p}{3u}\right)^3.$$

---

[5] We would group all forms together as $ax^3 + cx + d = 0$.

This can be rewritten as the quadratic

$$u^6 - qu^3 - \left(\frac{p}{3}\right)^3 = 0.$$

Since the quadratic was solved in antiquity, we have

$$u^3 = \frac{q}{2} + \sqrt{\left(\frac{q}{2}\right)^2 - \left(\frac{p}{3}\right)^3},$$

ignoring the negative (impossible) root.

Using (c), we come to

$$v^3 = -\frac{q}{2} + \sqrt{\left(\frac{q}{2}\right)^2 - \left(\frac{p}{3}\right)^3}.$$

And finally, $x = u - v$ gives us

$$x = \sqrt[3]{\frac{q}{2} + \sqrt{\left(\frac{q}{2}\right)^2 - \left(\frac{p}{3}\right)^3}} - \sqrt[3]{-\frac{q}{2} + \sqrt{\left(\frac{q}{2}\right)^2 - \left(\frac{p}{3}\right)^3}}.$$

Since our cube has side $a + b = u$, and $u = x + v$, we have solved the cubic by finding x. But $b = v$ so $x = a$. Thus we have solved the cubic by finding the small cube a^3.

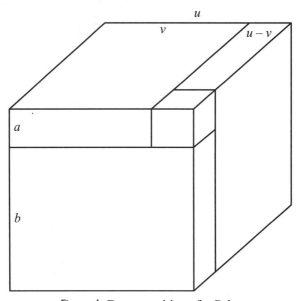

Figure 4. Decomposition of a Cube

How does this relate to our three-dimensional cube in Figure 4? The small cube in the upper right front corner is $(u - v)^3 = x^3 = a^3$, the larger cube in the lower back left corner (not visible in Figure 4, but visible in the lower left of Figure 5) is $v^3 = b^3$ and the whole cube is $u^3 = (a + b)^3$. The three identical rectangular blocks have dimension u by v by $u - v$ (or $a + b$ by b by a). See also Figure 6.

The case of $x^3 = px + q$ is handled similarly by letting $x = u + v$, $p = 3uv$ and $q = u^3 + v^3$.

So the cubic equation can be interpreted physically as a cube. Just as the solution to the quadratic is the smaller square that is found when we complete the square, the solution of the cubic is seen to be a smaller cube making up part of the larger cube. Thus we have completed the cube.

Why is a Square Square and a Cube Cubical?

I have always found that bringing in non-mathematical history really sparks the interest of those students in class with non-scientific interests, such as the language arts majors. For example, there is an interesting connection between Cardano and Shakespeare.[6]

The model

The wooden cube shown in Figure 5 depicts the three-dimensional analog of the cut-and-paste solution to the quadratic discussed in the first part of this chapter.

When assembled, the resulting cube (Figure 5) has dimensions of 9 cm. This is a nice hand-held size. You may choose to make yours larger to be more easily seen in the classroom. The individual pieces shown in Figure 6 have the following dimensions:

> 6 cm cube,
> 3 cm cube,
> three $3 \times 6 \times 9$ cm rectangular blocks.

The most direct way to make these elements is to cut them from appropriately sized pieces of wood using a jigsaw or table saw. Then sand the pieces until they fit together to form a seamless, or nearly so, cube.[7]

Figure 5. Three-dimensional cubic

Figure 6. Three-dimensional cubic pieces

---

[6] In 1573, Edward de Vere, the Sixteenth Earl of Oxford (1550–1604) sponsored an English translation and publication of Cardano's *Cardanus's Comfort*. This is a work of philosophy as opposed to mathematics. In *Cardanus's Comfort* is the following passage,

> What should we account of death to be resembled to anything better than sleep…But if thou compare death to long travel…there is nothing that doth better or more truly prophecy the end of life than when a man dreameth that he doth travel and wander into far countries.

If the ideas in this passage sound vaguely familiar, then you probably know your Shakespeare. Hamlet's "to be or not to be" soliloquy has the same message. Many scholars believe Edward de Vere was Shakespeare. Did Shakespeare get some of his material from Cardano? Perhaps.

In his edition, de Vere also included one of his own poems, of very Elizabethan style.
[*"Shakespeare" By another Name*, Anderson, Mark, Gotham Books, 2005, pp. 64–65.]

[7] This cube as well as the cycloid and brachistochrone in the chapter by V. Fredrick Rickey, were constructed by the late Cliff Long.

In the classroom

By observing the blocks depicted in Figures 5 and 6, you will see two cubes and three identical rectangular blocks. The dimensions, and thus the volumes, of these are: a^3, b^3, and $ab(a+b)$. When assembled, the new cube has dimension $a+b$ for a volume of $(a+b)^3$. Thus these cubes give visual confirmation of the fact that $(a+b)^3 = a^3 + 3ab(a+b) + b^3$.

Viewing the blocks in this light is the tack I most often take when using them in the classroom, whether it be a general history of mathematics class or a pre-calculus or calculus class. In these classes I do not go into the algebraic gymnastics presented above since in these courses the solution of the cubic is not covered, and might be too complex for the students. I give them the blocks and let them explore on their own. If this is done during work on completing the square, the three-dimensional version becomes a challenge problem. If it is done during a section on the cubic, this is a physical way of exploring the concepts. And if this is a lesson on the binomial theorem, it is a way for students to view the algebraic formula visually and geometrically. In these cases I do not present much more than what is represented by equations 1–3 above.

In a course that discusses the solution of the cubic in more detail, I would use the blocks first as a way to encourage students to try to find the solution to the cubic on their own. However, similar discovery exercises should be given in preparation for this. This is a fairly complex algebraic process and will be frustrating for the students if they are not accustomed to formulating solutions on their own. I use the blocks again as a visual aide as the solution is being presented to the class. This requires that time is set aside for a full look at the solution and its history. If you want to really side-track, ask what the solution of the quartic would look like!

Conclusion

The vast majority of students have to complete the square and use the quadratic formula by the time they reach college. By using the cutouts and presenting the historical background, the procedure and formula they are asked to memorize take on meaning that is otherwise lost to them. By going a step further and using blocks to present the three-dimensional analog, we encourage students to look past what is traditionally presented to try and foresee the continuation of the theory. A cubic equation actually represents a physical cube, and as such, can help guide them in understanding the solution.

Too often students view mathematics as a laundry list of unrelated and disconnected rules. By presenting the historical basis for some topics, especially the most common ones, we enable our students to view mathematics in a more holistic manner. We can encourage them to think of the connections in mathematics and share in the joy of discovering mathematical generalizations.

References

1. Berlinghoff, W. and Gouvea, F., *Math Through the Ages: A Gentle History for Teachers and Others*, Oxton House Publishers, 2002, pp. 105–108.
2. Calinger, Ronald, *A Contextual History of Mathematics*, Prentice-Hall, 1999, pp. 414–429.
3. ———, *Classics of Mathematics*, Prentice Hall, 1995, pp. 204–211, 261–266.
4. Mankiewicz, Richard, *The Story of Mathematics*, Princeton University Press, 2000, pp. 11, 78–81.

The Cycloid Pendulum Clock of Christiaan Huygens

Katherine Inouye Lau
Brown University

Kim Plofker
Union College

Introduction

The cycloid was an important "new curve" attracting mathematicians' attention in the seventeenth and eighteenth centuries. It turned out to be particularly significant in the study of the behavior of objects falling under the force of gravity: the cycloid is not only the brachistochrone (path of descent in shortest time) but also the tautochrone (path of descent in equal time from any point on the path). New mathematical tools such as the calculus made it possible to apply the study of such curves, and of concepts such as their "evolutes" and "involutes", to mechanical problems.

The significance of these developments is often lost on students who find them unfamiliar and remote. The story of Huygens' cycloid pendulum clock is an intriguing, easy-to-understand application of these mathematical ideas to a very practical problem. And it supplies a hands-on construction project that reinforces students' comprehension of how the cycloid and evolutes of curves actually work.

Huygens and the cycloid

Timekeeping problems and the tautochrone curve

In the middle of the seventeenth century, the scientific revolution and nautical discovery were in full swing. The expansion of trade and colonization meant an increasing need for accuracy in determining longitude at sea. An accurate clock would solve the problem of measuring time differences precisely enough to determine longitude; it would also be useful in many scientific experiments. The trouble was that clockmaking technology at that time wasn't developed enough to produce a sufficiently accurate clock. A "good" clock of the period might lose or gain as much as a quarter-hour per day. (The longitude problem and its influence on clockmaking have lately been the subject of several studies [1], [10].)

Clockwork mechanisms were not reliably regular: that is, they did not repeat their motions in equal time intervals, because the mechanical forces that drove them fluctuated too much. Around 1600, Galileo had recognized that the solution to this problem might lie in the physics of the ordinary pendulum [2]. The time a given pendulum took to swing in a circular arc appeared constant, independent of the amplitude of the swing. That meant that its oscillations would be constant—tautochronous—even if their amplitude changed. So the pendulum could be used for an accurate timekeeping device even though its oscillations grew smaller as it swung back and forth. (Galileo gave his own account of his pendulum experiments in *Dialogue Concerning Two New Sciences* [6].)

Huygens' clockwork

Unfortunately, the circular arc of a swinging pendulum is only approximately a tautochrone. The pendulum will in fact take longer to swing through large arcs than through small ones. This disadvantage was overcome by the Dutch mathematician Christiaan Huygens in his work on the pendulum clock.

The Netherlands in the seventeenth century was a major maritime power, so Dutch scientists were well aware of the practical importance of building better clocks. Huygens, born in 1629 into a wealthy family of the Netherlands aristocracy, soon showed an aptitude for both mathematics and mechanical tinkering [5]. In his publications from the 1650s till the end of his life in the 1690s, he investigated topics such as squaring the circle, telescope construction, the mechanics of elastic bodies, and the nature of light (his explanation of how wave propagation can produce the effects of geometrical optics is now known as "Huygens' principle"). The mechanical and mathematical problems involved in improving the performance of clocks also occupied much of his attention throughout his scientific career [4].

When Huygens initially invented the pendulum clock in 1656–57, he used an empirical solution to make the pendulum's swing more tautochronous [11]. Realizing that the circular arc needed to become steeper at its ends, he enclosed the string on which the pendulum bob hung between two metal "jaws" or "cheeks" whose lower ends curved outwards. Now the swinging string, instead of remaining straight like the radius of a circle, curved along the surface of the "cheek" and at the end of the swing lifted the bob above its former circular path. The steeper arc caused the bob to fall faster, counteracting the slowing-down effect of the larger swing. (See the illustration of Huygens' design in the chapter by V. Frederick Rickey in this volume [8].)

The mathematics of the cycloid and "evolving" curves

In 1659 Huygens began to analyze his empirical "fix" mathematically, trying to determine what sort of curve for the pendulum's swing would make it truly tautochronous. He found that the curved path he was looking for was surprisingly similar to another curve he had recently studied: the path traced out by a point on the circumference of a circle rolling along a straight line. Huygens was then able to prove that this curve (known as the cycloid) was indeed exactly the path that a truly tautochronous pendulum would follow.[1]

Huygens' proof, presented in his 1673 book *Horologium Oscillatorium* or *The Pendulum Clock* [3], was a lengthy geometrical demonstration (he considered geometry more mathematically sound than the newfangled methods of infinitesimals). The gist of the standard calculus proof of the tautochrone property of the cycloid is as follows.

Considering the bob to be falling in uniform gravity along some arc to the lowest point of its swing, the equations of motion give its velocity v only in terms of gravitational acceleration g and y, the height of the fall:

$$\frac{1}{2}mv^2 = mgy. \tag{1}$$

The velocity is ds/dt, the change in distance traveled along the arc with respect to the change in time:

$$\frac{ds}{dt} = \sqrt{2gy}. \tag{2}$$

If the arc of descent is chosen to be half a cycloid, then ds can be expressed in terms of the parametric equations for the cycloid, namely

$$x = a(\theta - \sin\theta), \quad y = a(1 - \cos\theta). \tag{3}$$

---

[1] See the above-cited chapter [8] for more detail on discoveries of the properties of the cycloid. An interesting discussion of how Huygens identified the cycloid in this problem, with some reproduced sketches from his own notebook, is presented in an online paper [7].

The Cycloid Pendulum Clock of Christiaan Huygens

We can solve for dt in terms of $d\theta$ and integrate it over the entire descent to get the time of descent. But if we now pick any other point on the same half-cycloid and substitute its value into the same equations instead, the expression for the time of descent turns out to be exactly the same. Thus the cycloid must indeed be the tautochrone curve.

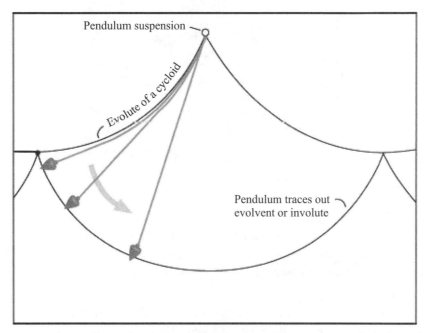

Figure 1. Evolute tracing out a secondary curve

Having found that the pendulum's path must be a cycloid, Huygens sought to know what shape the enclosing "cheeks" of the pendulum should be. What sort of curve would cause the end of the pendulum string to trace out a cycloid as it fell away from the curve? In fact, that curve is also a cycloid. This remarkable feature of "this marvelous line", as Huygens called it, led him to investigate the general relationship between a given curve (which he named an "evolute") and the secondary curve ("evolvent" or "involute") traced out when the first curve is straightened out starting at one end, like a pendulum string falling away from its "cheek". (Figure 1 shows this "evolution" in the case of the cycloid. The general properties of evolutes and involutes also proved useful in the design of the teeth of rotating gears, whose curves need to fit into and unroll from one another smoothly.)

Construction of the model

Using Huygens' idea

Although the complete mechanism of Huygens' cycloidal pendulum clocks is quite complicated, the cycloidal pendulum itself is much easier to build. Constructing the model shown in Figure 2 is an appropriate task for advanced high school and college students. The model compares the swing of a pendulum restrained by cycloidal "cheeks" to that of another pendulum oscillating freely in a circular arc. The regulating effect of the "cheeks" is surprisingly evident, even without any driving mechanism. Furthermore, the hands-on experience of building the model is a wonderful way to become very familiar with the shape and complexity of the cycloid.

For visual inspiration, students can look at woodcuts of Huygens' detailed clock mechanisms from his book [3].

Figure 2. Cycloid pendulum model

Step-by-step directions

For the construction, you will need one free afternoon, brief assistance from a friend, two pendulum bobs, a dowel half an inch in diameter, thin wire, a long straightedge, a sharp pencil, a large sheet of tracing paper, cardstock (two pieces 12 × 12 inches), and cardboard (a panel 18 × 24 inches and a circle seven inches in diameter).

- Prepare the tool for drawing the cycloid: a circle seven inches in diameter cut from cardboard, with a notch on the circumference for the point of a pencil.
- On a cardboard panel about 18 inches tall by 24 inches wide, draw a horizontal line which bisects the panel.
- Next, roll the circle along the underside of your horizontal, using a ruler as a guide (see Figure 3). Begin and end with the pencil on the horizontal line, about an inch from either vertical edge of the panel. With a friend's help to hold the pencil, use a continuous rolling motion to draw a smooth curve. You have created the **evolvent** or **involute** that will be traced out by the cycloidal pendulum.

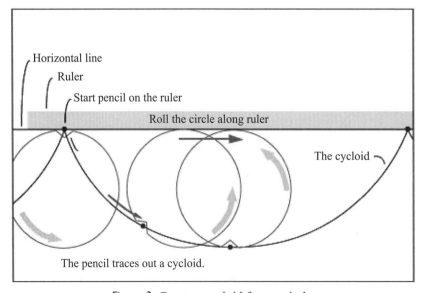

Figure 3. Create a cycloid from a circle

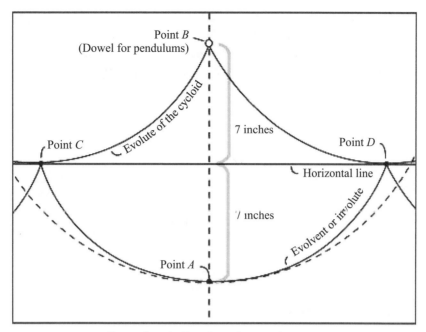

Figure 4. Labeling the points on the panel

- Trim the left or right edge of the panel so that the evolvent is centered.
- Label the lowest point on the evolvent *A* (see Figure 4). Seven inches above the horizontal and directly over *A*, mark the point from which the pendulums will hang as *B*. Label the two points where the evolvent intersects the horizontal as *C* and *D*.
- To create the corresponding evolutes, use tracing paper to transfer the curve *AD* to *CB* and then *AC* to *DB*.
- Compare the cycloid to a circular arc by drawing a circle with center *B* and radius *AB*, as in Figure 5. In the completed model each pendulum will attach at *B*, but one will follow a circular path while the other will be compelled to trace out the cycloidal path.

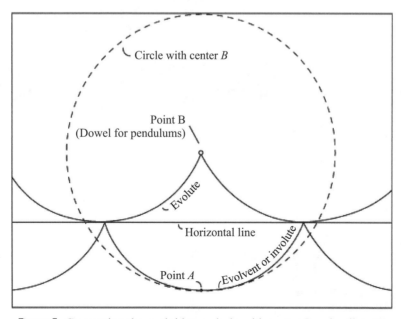

Figure 5. Comparing the cycloid to a circle with center *B* and radius *AB*

- Along the evolute, construct two "cheeks" as shown in Figure 2. Each rectangular strip juts out perpendicular to the panel, confining the pendulum between the evolutes to a cycloidal path. Each "cheek" is attached to the panel by a fringe along its length and a triangular tab at either end. Cut and fold the "cheek" from cardboard and the fringe from cardstock, using the template shown in Figure 6.

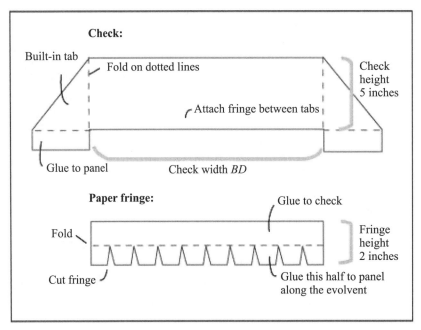

Figure 6. Template for cheeks, tab, and fringe

- Glue the uncut half of the fringe to the cheek and then attach both cheeks to the panel as shown in Figure 7. Align the cheek curve with the curve of the evolute.

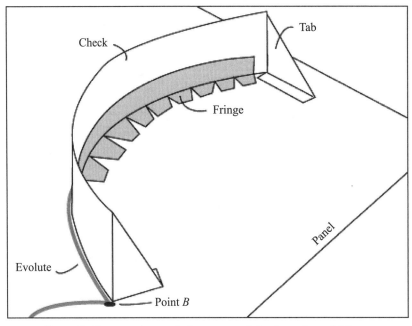

Figure 7. Construction and placement of cheeks, tab, and fringe

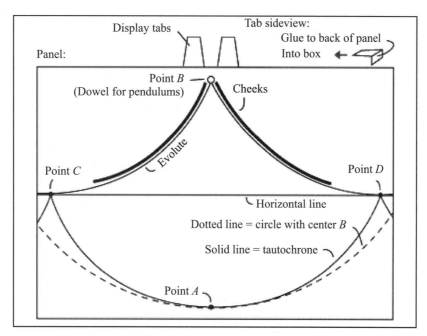

Figure 8. Adding the dowel and display tabs

- At point *B*, cut a hole and insert a sturdy dowel (half an inch in diameter), as shown in Figure 8.
- For display, tabs glued to the top of the panel can tuck under books on a shelf or into the top fold of a weighted cardboard box. For stability, insert the dowel completely through the panel and the box.
- Finally, attach to the dowel two small pendulum bobs (available at hardware stores) suspended by thin wire or inelastic string. Install the first bob close to the panel where its string will curve against the cheeks and produce a cycloidal path. Position the second bob farther from the panel, where it will swing freely in a circular path. (If the pendulums tend to slide along the dowel, skewer cardboard tabs between them as shown in Figure 2.)
- Pull up both pendulum bobs to the same height; release them and observe the difference in their swings.

Conclusion: possibilities for further investigation

Students at the advanced high school or college level, with some calculus training, can explore this project in a class on the history of mathematics (particularly the historical development of calculus) or history of science. It would serve as a useful demonstration tool for a student presentation on the topic of evolutes and involutes or the tautochrone. Pendulum clocks in general are a fascinating introduction to much of the mathematics and mechanics of the seventeenth century; a student inspired to learn more about them might like to seek out physical examples somewhat more complex than this cardboard model. A reproduction of Huygens' clock at the Science Museum in London can be seen online [9].

Acknowledgements

The historical and mathematical description of Huygens' work on the cycloid partly relied on a paper written by a fellow student of the first author, Kate Magaram. The authors are also indebted for several helpful comments to the anonymous reviewers of the manuscript and to Professor Henk Bos of the University of Utrecht.

References

1. Andrewes, William J. H. (ed.), *The Quest for Longitude*, Harvard Collection of Historical Scientific Instruments, Cambridge (MA, USA), 1996.
2. Bedini, Silvio A., *The Pulse of Time: Galileo Galilei, the Determination of Longitude, and the Pendulum Clock*, Olschki, Florence, 1991.
3. Blackwell, Richard J. (trans.), *Christiaan Huygens' The Pendulum Clock or Geometrical Demonstrations Concerning the Motion of Pendula as Applied to Clocks*, Iowa State University Press, Ames (Iowa, USA), 1986.
4. Bos, H. J. M., "Christiaan Huygens", in C. C. Gillispie (ed.), *Dictionary of Scientific Biography*, vol. 6, Charles Scribner's Sons, New York, 1972, pp. 597–613.
5. Bos, Henk J., "Christiaan Huygens", in *Lectures in the History of Mathematics* American Mathematical Society/London Mathematical Society, Providence (RI, USA), 1993, pp. 59–82.
6. Galileo, *Dialogues Concerning Two New Sciences*, trans. Henry Crew and Alfonso di Salvio, Dover, New York, 1954.
7. Mahoney, Michael, "Sketching science in the seventeenth century", http:// www.princeton.edu/~hos/mike/articles/whysketch/whysketch.html}
8. Rickey, V. Frederick, "Build a Brachistochrone and Captivate Your Class", in Amy Shell-Gellasch (ed.), *Hands On History: A Resource for Teaching Mathematics*.
9. Science Museum in London, Huygens' clock reproduction: http://www.sciencemuseum.org.uk/objects/time_measurement/1956-220.aspx
10. Sobel, Dava, *Longitude: the true story of a lone genius who solved the greatest scientific problem of his time*}, Walker and Co., New York, 1995, revised ed. with illustrations, Dava Sobel and William J. H. Andrewes, *The Illustrated Longitude*, Walker and Co., New York, 1998.
11. Vermij, Rienk, et al., *Christiaan Huygens* (in Dutch), Epsilon Uitgaven, Utrecht, 2004.

Build a Brachistochrone and Captivate Your Class

V. Frederick Rickey
United States Military Academy

Introduction

Cliff Long (1931–2002) [5] was a master teacher whose office was a wonderful place to visit, for it was crammed with a wealth of teaching devices. From his early *Bug on a Band* [6], to his slides and flexible model of quadratic surfaces [3,4], to his head of Abraham Lincoln made with a computer-controlled milling machine [1], and his fascination with knots [7], Cliff was always on the lookout for new ways to illustrate mathematical concepts.

As a young faculty member I went to his office whenever I wondered how best to present some topic in class. He had thought long and hard about everything he taught and was full of ideas about how to enhance learning. Cliff was my mentor and I learned an immense amount about teaching from him.

Of all the things in his office, my favorite was his brachistochrone. I borrowed it to use in talks whenever the Bernoullis were mentioned. The brachistochrone problem was my favorite way to end a class on the integral calculus, for it provided a lovely way to review many of the topics we had studied [10]. Shortly before I retired from Bowling Green State University in 1998, Cliff talked to me about an improved design for the brachistochrone and asked for my suggestions. Little did I know that he was making one for me. I was honored.

Parametric equations

When introducing the topic of parametric equations, a good way to proceed is to arrive in the classroom with your brachistochrone under your arm. When your students arrive, take the circular disk and draw a cycloid on the blackboard (see Figure 1). Be sure to tell the students that a cycloid is the curve generated by a point on the circumfrence of a circle that is rolling along a straight line and that the word derives from the Greek word for circle, *kuklos*. Perhaps they will have heard of epicycloids, those circles rolling on

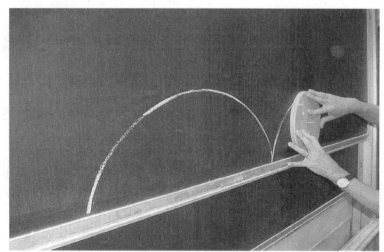

Figure 1. Rolling a curve to produce a cycloid

circles, that are part of Ptolemaic astronomy. Of course you will want to practice drawing a cycloid in advance, for it is a little tricky to get the disk to roll without slipping; also you need to keep enough pressure on the disk—and the chalk—so that the chalk will trace the curve. If the blackboard does not have a protruding edge that you can roll the disk on then you will have to get several students to hold a meter stick tightly against the blackboard so that you will have a firm base to roll the circle on. If the blackboard slides up so that you can work at a more convenient height, that is even better.

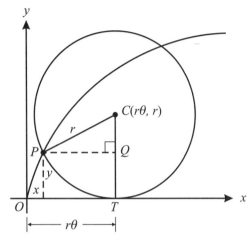

Figure 2. Finding parametric equations for the cycloid

Now we are ready to derive parametric equations for the cycloid. If you place the wooden disk on the base line and then trace around it with chalk, you can draw a circle on the base line of your cycloid, as in Figure 2.

For the parameter we use θ, the angle the circle has rolled through. We begin with $\theta = 0$ and the tracing point P at the origin. If the circle has radius r, then when it rolls through angle θ the circle will roll the distance $OT = r\theta$. To locate the x-coordinate of the point P, first move to the right the distance OT and then left the distance QP, i.e., $r\sin\theta$. Similarly, to obtain the y-coordinate we move up to the center of the circle (TC), i.e., up r. and then down to P, i.e., down CQ or $r\cos\theta$. Thus we have the parametric equations. Moving left (down) accounts for the minus sign in the parametric equations.

$$\begin{cases} x = r\theta - r\sin\theta, \\ y = r - r\cos\theta. \end{cases}$$

At this stage your students should plot these equations on a computer algebra system and observe that the graph they obtain looks like what was sketched on the blackboard by the rolling circle.

While the early history of the cycloid is unknown, Galileo named the curve (it is also called the roulette and trochoid) and attemped to find the area under the curve in 1599. He approximated the area by making a physical model and weighing it. His conclusion was that the area under one arch of the cycloid was about three times the area of the generating circle. One could repeat Galileo's experiment in class by constructing an accurate arch of a cycloid and weighing it—but you will need a chemist to loan you an accurate scale. In 1643, Roberval proved Galileo's conjectured value of 3 was correct. Tangents to the curve were constructed later in the decade by Descartes, Fermat and Roberval, presumably independently. In 1658, Pascal posed a number of problems related to the cycloid. The most interesting of these, the rectification, or arc length, of the curve, was solved by Christopher Wren [17]. In 1686, Leibniz found a Cartesian equation for the curve:

$$y = \sqrt{2x - xx} + \int \frac{dx}{\sqrt{2x - xx}}.$$

Bernoulli's New Year's Day problem

Just nine months after Newton left the Lucasian Professorship at Cambridge to take up "y^e Kings business" at the Mint in London, he received a letter from France containing a flysheet printed at Groningen which was dated January 1, 1697. He received it on January 29, 1696; this is no typo, England was not yet on the Gregorian calendar [17]. It was addressed

> To the sharpest mathematicians now flourishing throughout the world, greetings from Johann Bernoulli, Professor of Mathematics.[11]

You might wonder why Johann Bernoulli was teaching in the Netherlands. This was because his older brother, Jacob, held the professorship of mathematics at the university in their native town of Basel; the newly married Johann was forced to look elsewhere. Through the help of Liebniz and L'Hospital he obtained a position at the university in Groningen. Bernoulli's stated aim in proposing this problem sounds admirable:

> We are well assured that there is scarcely anything more calculated to rouse noble minds to attempt work conductive to the increase of knowledge than the setting of problems at once difficult and useful, by the solving of which they may attain to personal fame as it were by a specially unique way, and raise for themselves enduring monuments with postcrity. For this reason, I ... propose to the most eminent analysts of this age, some problem, by means of which, as though by a touchstone, they might test their own methods, apply their powers, and share with me anything they discovered, in order that each might thereupon receive his due meed of credit when I publically announced the fact. [11]

Bernoulli's new year's present to the mathematical world was a great gift, a difficult problem that would enrich the field:

> To determine the curved line joining two given points, situated at different distances from the horizontal and not in the same vertical line, along which a mobile body, running down by its own weight and starting to move from the upper point, will descend most quickly to the lowest point. [11]

This is the brachistochrone problem. The word was coined by Johann Bernoulli from the Greek words 'brachistos' meaning shortest and 'chronos' meaning time. But the problem was not new. In 1638, Galileo attacked it in his last work, *Discorsi e dimostrazioni matematiche, intorno à due nuoue scienze* [*Discourses and Mathematical Demonstrations Concerning Two New Sciences*], but he was unable to solve it. Galileo was only able to prove that a circular arc provided a quicker descent than a straight line. Bernoulli noted this in his flysheet when he wrote that the solution to the brachistochrone problem was not a straight line, but a curve well known to geometers.

Earlier, in June 1696, Johann Bernoulli published a paper in Germany's first scientific periodical, the *Acta eruditorum*, wherein he attempted to show that the calculus was necessary and sufficient to fill the gaps in classical geometry. At the end of the paper the brachistochrone problem was posed as a challenge, setting a deadline in six months, but Bernoulli received no correct solutions. He had received a letter from Leibniz sharing that "The problem attacked me like the apple did Eve in Paradise" [14] and that he had solved it in one evening. In fact, he had only found the differential equation describing the curve, but had not recognized the curve as an inverted cycloid. Bernoulli and Leibniz interpreted Newton's six month silence to mean the problem had baffled him—indeed he had not seen it. To demonstrate the superiority of their methods, Leibniz suggested the deadline be extended to Easter and that the problem be distributed more widely. So Bernoulli added a second problem, had a flysheet published, and made sure it circulated widely.

The brachistochrone problem was a difficult one. Pierre Varignon and the Marquis de L'Hospital, in France, and John Wallis and David Gregory, in England, were all stumped. But Newton was not. Thirty years later Newton's niece Catherine Barton Conduitt recalled,

> When the problem in 1697 was sent by Bernoulli—Sr. I. N. was in the midst of the hurry of the great recoinage [and] did not come home till four from the Tower very much tired, but did not sleep till he had solved it wch was by 4 in the morning. [15, pp. 582–3; 16, pp. 72–73]

The next day Newton sent his solution to his old Cambridge friend Charles Montague, who was then President of the Royal Society. He published Newton's work anonymously in the February issue of the

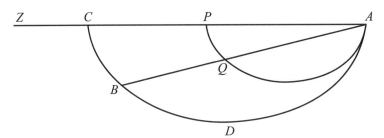

Figure 3. Newton's Construction

Transactions of the Royal Society of London. The trap that Bernoulli and Leibniz had set for Newton failed to snare its game.

Newton provided no justification for his solution, only showing how to construct the necessary cycloid [11, p. 22; 16, p. 75]. He simply drew an arbitrary inverted cycloid with its starting point at *A*, the higher of the two points, and then drew a line between the two given points, *A* and *B*. This line intersects the arbitrary cycloid at *Q*. Then he found the ratio of the line segment from the starting point to the final point to the line segment from the starting point to the initial cycloid, i.e., *AB/AQ*. He used this ratio to expand the radius of the initial circle to produce the circle which would generate the desired cycloid.

Derek T. Whiteside, who has published the extremely valuable edition of *The Mathematical Papers of Isaac Newton*, claims that the fact that it took Newton twelve hours to solve these problems indicates that his mathematics was rusty from nine months disuse. It also shows that the gradual decline in Newton's mathematical ability had set in. However, his solution of the brachistochrone problem is counterevidence to the myth that Newton's old age was mathematically barren [16, pp. xii, 3].

Immediately on receiving the solution of the anonymous Englishman via Basnage de Beauval, Bernoulli wrote Leibniz that he was "firmly confident" that the author was Newton. Leibniz was more cautious on the authorship of the problem, admitting only that the solution was suspiciously Newtonian. Several months later Bernoulli wrote de Beauval that "we know indubitably that the author is the celebrated Mr. Newton; and, besides, it were enough to understand so by this sample, ex ungue Leonem." Within a few weeks this shrewd guess was common knowledge across Europe. The phrase goes back to Plutarch and Lucian, who allude to the ability of the sculptor Phidias to determine the size of a lion given only its severed paw.

Not having succeeded in trapping Newton, Bernoulli lost interest, leaving it to Leibniz to publish the solutions in the May 1697, *Acta* (pp. 201–224). These included Johann's own solution, one by his older brother Jakob, one by L'Hospital (probably produced with the help of Johann Bernoulli), one by Tschirnhaus, and a reprint—seven lines in all—of Newton's. This time Newton was not anonymous, for Leibniz had mentioned him in his introductory note. Leibniz was so embarrassed by the whole thing that he wrote the Royal Society indicating that he was not the author of the challenge problems. Technically, this was true, but he had contrived with Bernoulli to embarrass Newton.

Within a few years there were solutions by John Craige, David Gregory, Richard Sault and Fatio de Duillier. In 1704 Charles Hayes, in his widely read *Treatise on Fluxions*, presented it as a mere worked example in this textbook. As often happens, a difficult problem, once cleverly solved, comes within the grasp of many.

Perhaps the most important of all of these solutions is that by Jakob Bernoulli. While somewhat ponderous, it led to a new field of mathematics, the calculus of variations, a branch of analysis where the variables are not numbers but functions.

The clever elementary solution of Johann Bernoulli is the one that I like to present at the end of a course on the integral calculus, for it provides both a wonderful story and a great review. You can find it in many places, e.g., [10, 12]. Translations of the solutions of the Bernoulli brothers can be found in [13].

Building your brachistochrone

One of my students, Zachary Seidel, had a brachistochrone shaped sliding board as a young child. His father had studied mathematics as an undergraduate and decided that his son should have the quickest slide in the neighborhood. Such a slide is probably too big for your classroom, so the model designed by Cliff Long will be described.

You will need a sheet of 3/4 inch plywood that measures roughly 30 by 14 in. First cut a strip off the long edge about 1 1/4 inches wide; its use will be described momentarily. From another piece of plywood, cut out a circle of radius 4.5 in. Drill a hole in the center and insert a 1/4 inch dowel that sticks up far enough to provide a nice handle. Near the circumfrence of the disk drill another hole that will hold a small piece of chalk (ideally the chalk would be *at* the circumference, but that makes the construction of our bachistochrone harder). The distance from the center of this hole to the circumference of the disk should be slightly more than the radius of the chalk. This is so the chalk will not fall out. You should also drill another hole in the disk at about half the radius so the chalk can be moved there later to draw a curtate cycloid.

Figure 4. The Finished Brachistochrone

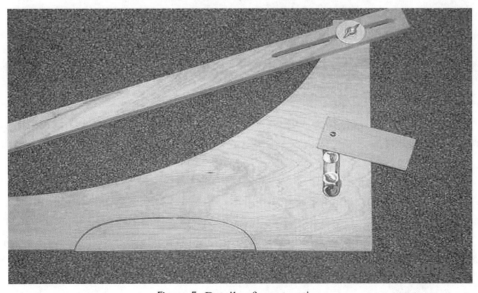

Figure 5. Details of construction.

Use this disk and chalk to trace out one arch of the cycloid on the sheet of plywood. Start with the chalk about 2 inches from the short edge and on the same side where you cut out the disk. Now cut on this chalk line and you will have a cycloid. Use your router to cut out a groove on the edge of the plywood so you will have a track for a marble to run down.

The fun part of this model is a straight line track for another marble to run down. It will be used to demonstrate clearly that the straight line is not the quickest path. To do this, you will use the piece of plywood that you cut off earlier. Cut a groove on one side for the marble. Cut a slot along the middle of this board starting about one inch from the end and about 12 inches long. The slot should be wide enough so that it will slide over a set screw. The idea is that you want to adjust where the straight line track ends. At one extreme it should be possible to go almost horizontally across the whole arch of the cycloid; on the other, you want to be able to adjust it so that the straight line track ends at the low point on the inverted cycloid, and perhaps even before that.

Next cut out a semi-elliptical piece on the yet untouched straight base. With a screw, and washer used as a spacer, this will provide a toggle stand for your brachistochrone. Finally, you need a place to store several marbles (three are suggested in case you lose some of your marbles in class). See Figure 5.

Using your brachistochrone in class

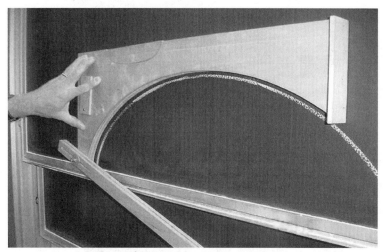

Figure 6. Brachistochrone = Cycloid.

After you have drawn a cycloid on the chalkboard, and before you have derived its equation, hold your brachistochrone up to the chalkboard to show the students that the curve you just drew is the same curve as your brachistochrone. Now you can explain Johann Bernoulli's problem and let them test the cycloid solution against the straight line solution.

This is an opportunity to get your students involved. Invite several of them to adjust the straight line chute so that it ends at the low point of the inverted cycloid on the brachistochrone. Have one of them get out two marbles and hold them with one hand so that one marble is on the brachistochrone track and the other is on the straight line track. By using one hand it is easier to release the marbles simultaneously. Some chaos will result when the marbles go flying off the end of the tracks and across the room. Probably some students will be disbelievers—when one first encounters this demonstration it is hard to believe that the cycloid track is quicker than the straight line track—and want to try themselves. This is good; get everyone involved.

Next adjust the straight chute to direct one marble to another point. Pick a point that is beyond the low point of the brachistochrone, on the uphill slope. This time, to the astonishment of the students, the marble race is even more unfair. Travel along the brachistochrone is much quicker. Students are perplexed that the quickest way to get from A to B is to dip below B and then coast back up to it. They will ask why this happens. They do not realize that they are asking to see the proof that the brachistochrone really is the quickest path. This shows the motivational power of interesting historical examples combined with a tactile demonstration of the result.

I have met mathematicians who believed that to get from A to B on the quickest path, that one should draw a cycloid that begins at A and has its low point at B. To see that this is incorrect, examine Newton's picture again (Figure 3). A little thought will show that it is rarely the case that B is at the bottom of the cycloid. If the points are far apart, it is impossible to get from the upper to the lower point without dipping below the lower point and then coasting back up.

The brachistochrone model can be used to demonstrate another interesting property of the cycloid. The curve is isochronous, i.e., no matter where one starts on the curve the time it takes to reach the low point is the same (see [2] for the original proof). To illustrate this, have a student hold a marble in each hand, one on each side of the minimum point on the cycloid, but at different distances from the midpoint, and then release them simultaneously. They will collide at the low point of the cycloid. To confirm this the observer needs to be directly in front of the model and to focus on the minimum point of the cycloid (a mark on the model will help locate this point for the observer). Your students will want to repeat this multiple times from different starting positions thereby giving all a chance to observe up close. Andy Long, Cliff's son, suggests another way to do this. Ask your students to close their eyes and remain quiet so that they can hear the equal beats of a single marble rolling back and forth on the cycloidal track. Use a large amplitude so that it goes back and forth a number of times. Your students will unconsciously rock their heads back and forth in time with this cycloidal clock.

The ordinary pendulum is not isochronous; the period T depends upon both the length L and the angle of oscillation θ. This provides an interesting real world example of a multivariable function, which is due to Daniel Bernoulli in 1749.

$$T(L,\theta) = 2\pi \sqrt{\frac{L}{g}}\left(1 + \frac{1^2}{2^2}\sin^2\frac{\theta}{2} + \frac{1^2 + 3^2}{2^2 + 4^2}\sin^4\frac{\theta}{2} + \cdots\right),$$

When the angle of oscillation is small, then all of the terms of the series involving the sine can be ignored. This provides yet another point in the curriculum to bring in this circle of ideas.

Christiaan Huygens, the foremost mathematical physicist in the generation before Newton, took advantage of this property of the cycloid to design an accurate pendulum clock (Figure 7). But he needed one more mathematical property of the cycloid: the evolute of a cycloid is another cycloid of the same size.

He hung the bob of his pendulum on a thread that swung between cycloidal cheeks. The cheeks prevented the bob from swinging in a circular arc like in a regular pendulum. When the bob moved to the side it wound around the cycloidal cheeks and was pulled up slightly. He showed that the curve of the bob was a cycloid.

Huygens published this work in his *Horologium oscillatorium* of 1673; you can read how he constructed a cycloid (he had a clever device for avoiding slippage) and designed his clock [2]. A model of his clock can be seen in the Borehaave, a wonderful science museum in Leiden in the Netherlands. Unfortunately, as one might suspect,

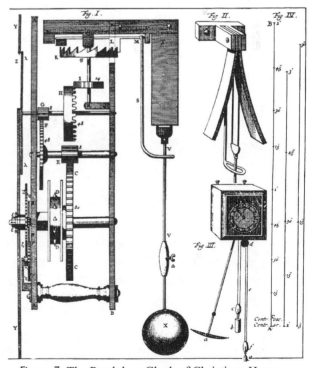

Figure 7. The Pendulum Clock of Christiaan Huygens

the accuracy of Huygens' clock was killed by friction. This illustrates the old adage: In theory, theory and practice are the same, but in practice, theory and practice are not the same.

Trains and epicycloids

A rewarding question to pose to students is this: As you watch a speeding train go by, what point on it is moving backwards? It will probably take some prompting to get them to realize that the answer is the bottom of each wheel, the flange, which is the part of the wheel below the top of the rail and which keeps the train on the track. This will give us another opportunity to use our device, but we will need another piece to attach to our cycloid drawer, a piece of wood several inches longer than the radius of our disk. It should have a hole which goes over the peg at the center of the disk, another peg to go in the hole where the chalk went while drawing the cycloid, and another hole near the end for another piece of chalk. This arrangement forces the board to turn as the disk turns. If the chalk is at distance $d > r$ from the center of the circle, then it produces a prolate cycloid. When you use this device to draw a prolate cycloid in class, your students will literally be able to see that when the chalk is at the bottom of the curve it is moving backwards.

The student who understands how the parametric equations for the ordinary cycloid were obtained will have no trouble finding the equations for a prolate cycloid:

$$\begin{cases} x = r\theta - d\sin\theta, \\ y = r - d\cos\theta. \end{cases}$$

If we compute the derivative in the x-direction we obtain

$$\frac{dx}{d\theta} = y = r - d\cos\theta$$

If θ is (near) an even multiple of π, then this derivative is negative. In practical terms, the bottom of the train wheel is moving backwards.

When d is arbitrary the curve is called a trochoid, when $d > r$ it is a prolate cycloid and when $d < r$ it is a curtate cycloid. Curtate cycloids are used by some violin makers for the back arches of some instruments, and they resemble those found in some of the great Cremonese instruments of the early 18th century, such as those by Stradivari [8].

Conclusion

One does not have to prove everything one talks about. We mathematicians need to start talking about mathematics. There are so many interesting things in mathematics that will captivate students. Once learners are interested they will be motivated to get a deeper understanding of that mathematics. They will seek out the proofs on their own.

We have seen the essential ingredients of good teaching combined in one problem. It is a problem with a fascinating history, connected to a multitude of important names that our students should know. The mathematics is tractable to undergraduates. The ideas can be used at several places in the curriculum and this spiral approach enhances learning. Finally one has a nice classroom device. For all of this, I thank Cliff Long.

The author would like to thank Andy Long of Northern Kentucky University, Tom Hern of Bowling Green State University, and Florence Fasanelli of AAAS for helpful commments on drafts of this paper and colleague LTC Michael Huber for "lending a hand" in the preparation of this paper.

References

1. Thomas Hern, Cliff Long and Andy Long, "Looking at Order of Integration and a Minimal Surface," *The College Mathematics Journal*, 29 (1998) 128–133.

2. Christiaan Huygens, *Horologium oscillatorium* (1673). English translation as *Christiaan Huygens' the pendulum clock, or, Geometrical demonstrations concerning the motion of pendula as applied to clocks*, translated with notes by Richard J. Blackwell; introduction by H. J. M. Bos, Ames: Iowa State University Press, 1986.

3. Cliff Long, "Peaks, Ridge, Passes, Valley and Pits: A Slide Study of $f(x, y) = Ax^2 + By^2$," *The American Mathematical Monthly*, 83 (1976) 370–371.

4. ———, "A Flexible Model for Peak, Ridge and Pass," *The College Mathematics Journal*, 7 (1976) 16–17.

5. ———, 1931–2002, http://www.bgsu.edu/departments/math/Ohio-section/CliffLong/

6. ———, *Bug on a [Möbius] Band*: http://www.wcnet.org/~clong/carving/bug.GIF

7. ———, The Frenet frame. http://www.wcnet.org/~clong/carving/3knotsandtwist.GIF

8. (Playfair 1990). http://mathworld.wolfram.com/CurtateCycloid.html

9. Rickey, V. Frederick, "Mathematics of the Gregorian calendar," *The Mathematical Intelligencer*, 7(1985), no. 1, 53–56.

10. ———, "History the Brachistochrone," to appear in *Historical Notes for Calculus Teachers*. Available at http://www.dean.usma.edu/departments/math/people/rickey/hm/CalcNotes/brachistochrone.pdf

11. J. F. Scott, ed. *The Correspondence of Isaac Newton*, Vol. IV, 1694–1709, Cambridge University Press, 1967. See letters #561 to Montague on 30 Jan 1969/7 (pp. 220–229) and #601 to Flamsteed 9 Jan 1698/9 (296–7), which are in English and Latin.

12. George F. Simmons, *Differential Equations with Applications and Historical Notes*, McGraw-Hill, 1972. See pp. 25–31.

13. Dirk J. Struik, *A Source Book in Mathematics, 1200–1800*, Harvard University Press, 1969. Contains English translations of the solutions of the brachistochrone problem by Johann and Jakob Bernoulli, pp. 391–399.

14. Rüdiger Thiele, "The brachistochrone problem and its sequels," unpublished manuscript of a lecture given at the Frederik Pohle Colloquium, Adelphi University, January 24, 2001.

15. Richard S. Westfall, *Never at Rest. A Biography of Isaac Newton*, Cambridge University Press, 1980. See pp. 581–583.

16. D. T. Whiteside, *The Mathematical Papers of Isaac Newton, Volume VIII, 1697–1722*, Cambridge University Press, 1981. Contains excellent and very scholarly commentary by Whiteside (pp. 1–14) as well as the pertinent documents (pp. 72–90) and numerous references.

17. E. A. Whitman, "Some historical notes on the cycloid," *The American Mathematical Monthly*, 50 (1943), 309–315. Reprinted in *Sherlock Holmes in Babylon and Other Tales of Mathematical History*, MAA, 2004, pp. 183–187.

Exhibiting Mathematical Objects: Making Sense of your Department's Material Culture

Peggy Aldrich Kidwell
National Museum of American History, Smithsonian Institution

Amy Ackerberg-Hastings
University of Maryland University College

Introduction

Most of the chapters in this volume suggest ways to use historical mathematical instruments or replicas of those instruments to highlight mathematical or pedagogical principles within the classroom. Yet, some teachers and professors may wish to bring these objects to a wider audience. An on-line or physical exhibit is one venue for increasing public awareness of mathematics and of one's own mathematics department. This chapter outlines fundamental principles of exhibit planning to help professors, teachers, and students identify, understand, and arrange historical objects and books that might be available to them. It suggests methods appropriate to a range of projects, from those displayed for a single day to cases professionally designed at the time of a major school anniversary or building renovation. It is illustrated with examples of models, devices, and books held by the Smithsonian's National Museum of American History (NMAH) and the Smithsonian Institution Libraries.

Preliminaries

What is meant by the "material culture" of mathematics? This term simply refers to the objects used in mathematical research and teaching. It includes books, letters, and manuscript notes relating to mathematics. It encompasses drawing instruments as well as computing devices such as slide rules, planimeters, adding machines, and electronic calculators. It also includes physical objects that teach mathematical principles—certain games, toys and puzzles; geometric models; and mathematically important software. If you, your colleagues, or your students have any of this mathematical "stuff" tucked away in a drawer, closet or library, you have the raw materials for an exhibit. This chapter suggests how you might organize such materials into a coherent presentation.

Why exert the effort to prepare a mathematics exhibit? There are several potential benefits. It is fun to learn about what these objects are, who used them, and how they were used. This new knowledge will enhance your own sense of mathematics and its past. Some of this knowledge and enjoyment may rub off on your visitors. An exhibit that contains objects that were once used—or are like those once used—at your institution also contributes to a broader historical account of the school and encourages visitors to explore that history. An exhibit might serve as a group project in a mathematics or history of mathematics class.

Figure 1. "Slates, Slide Rules and Software: Teaching Math in America."
Negative number 2002-1655. Courtesy of the Smithsonian's National Museum of American History.

What if you do not have any objects? First, look around your library, your classrooms, your office, and your home. Is there an old teaching slide rule? A blackboard made of slate? Are there strange looking string, plaster, paper, or plastic models? Early commercial graphing calculators or software packages? Published journals with papers by Einstein or anyone else whom you would like to discuss? Mathematical puzzles and toys? Does another department such as physics or engineering have punch cards, calculating machines, planimeters, drawing instruments, or slide rules? Is there an institutional museum? Does someone in your department make models, fold origami, or collect mathematical mouse pads? Does an alumnus have a collection? Is there a sundial on campus? Did the school newspaper or yearbook show anything about math students, faculty or clubs? How, if at all, does your school show up in a search of the mathematical journals preserved on databases such as JSTOR?

Remember, today's college students were born in the late twentieth century. What is familiar to you may be unknown to them and vice versa. If there is truly nothing on campus, consult your local historical society and/or public library. They may have intriguing books and objects they would love to put on display. Of course, you might also want to talk about objects far older than anything your school has. If so, you could combine photographs of the real thing with clearly identified replicas. Indeed, making the replica could be the subject of the exhibit. Here you might want to talk about what you think is *mathematically* essential to representing a given type of object (e.g., using the correct projection on an astrolabe) and what is not (making it out of brass). That is, you might want to try to say what a replica does and does not represent about the historical object it attempts to duplicate.

Where could an exhibit be mounted? Traditionally, objects have been shown in glass cases (preferably, closed locked cases with alarms). Older buildings at schools and colleges often have empty display cases in hallways near entry doors or classrooms. The keepers of those cases might be pleased to have someone volunteer to fill one of them. Campus librarians also often mount changing exhibits and might be amenable to a mathematics display. There is information about constructing and installing new cases in *Good Show*. [70] If you have students working on an exhibit, you can also encourage them to prepare a poster or tabletop exhibit for presentation at a conference. Finally, virtual exhibits are increasingly appearing on the World Wide Web. Several on-line exhibits are listed in the references and can serve as models.

How can someone without curatorial experience create an appealing exhibit? In the remainder of the chapter, we will outline some fundamental and uncomplicated steps that amateurs can learn from the professionals. These steps will be illustrated with two exhibits prepared by the authors. The first was "Slates, Slide Rules and Software: Teaching Math in America," a showcase curated by Peggy Kidwell, the first author of this chapter, and built by Smithsonian staff over a period of several months. It had a budget of several thousand dollars and was on display at NMAH from February 2001 to February 2004 (Figure 1). It lives on as an on-line exhibit. [24] The second was "What Every Math Major Needs to Know," an exhibit of rare mathematics books from the collections of the Dibner Library at NMAH that Amy Ackerberg-Hastings, the second author of this chapter, developed over a period of several weeks and with no allotted budget (Figure 2). It was on view during a few days of July 2002.

We hope that by reading about what went into these exhibits, you will gain a better sense of what you and your students might do. Much of what we will say about selecting an organizing scheme, finding out about objects, and combining visual and written information would apply even if you choose to present your ideas in a virtual exhibit by designing Web pages. Finally, these comments should make it easier if you decide to hire people with special expertise in areas such as exhibit design, object conservation, photography, and exhibit construction to work on your show.

Choosing an ordering scheme

An "ordering scheme" is the unifying structure of an exhibit. As is true of any writing project, you need to begin by writing a preliminary thesis statement. In other words, your first task is the intellectual exercise of

Figure 2. First of three cases for "What Every Math Major Needs to Know."
Negative number 2005-27185. Courtesy of the Dibner Library, Smithsonian Institution Libraries.

deciding what you are exhibiting and why you think other people ought to see these objects. For example, Peggy began with the thesis that mathematics teaching in the United States has reflected other aspects of American history. This led her to organize objects into four historical periods:

- The Early Republic (the period in the early nineteenth century when public elementary schools were established to educate citizens).
- The World Stage (the time of American economic prosperity around 1900 that saw the emergence of a research community in mathematics and the expansion of public high schools).
- The Cold War (the "New Math" era of the 1960s).
- The Information Age (the final decades of the twentieth century, when electronic calculators, microcomputers, and mathematical software became common).

As she dug further into this story, Peggy modified her thesis. Not only historical events, but changing demographics had a major influence on mathematics teaching. When primary, university, secondary, and, finally, preschool education expanded, or when the sheer number of schoolchildren grew, mathematics education changed at the same time. That is to say, Peggy did not develop a thesis or study her objects in isolation from each other. Rather, she went back and forth between objects and the ideas she had about them. Delving into a topic in this way is part of what makes an exhibit intellectually challenging.

This give and take process helped Peggy decide which objects and graphics she used in and collected for the exhibit. For instance, to represent relatively recent technologies, she acquired user manuals for the 1989 version of DERIVE™ software from Soft Warehouse, Inc., in Honolulu. To liven up the show, she found at least one image of people using objects for each section of the exhibit. For example, a photograph from Wellesley College shows students with their instructor examining models like those from NMAH's collections. Having an organizing scheme also helped rule out some objects. The Smithsonian has a device for teaching elementary arithmetic called a Visigraph that dates from about 1910. The Visigraph stands about 3 feet high. Including it would have dwarfed the objects relating to high schools and universities which she wished to highlight in that time period.

The periodization scheme also helped the designer of "Slates, Slide Rules." He divided the case into four units reflecting the four themes. (See Figure 1.) Each area had labels on a different color of board (white, lilac, green, and blue). The back of the case also was made from a material that represented the era (clapboard, brick, ceramic tile, and aluminum). These visuals helped guide visitors through the array of objects on exhibit.

For her project, Amy began by contemplating who would be viewing the exhibit: the sixteen participants in the George Washington University Summer Program for Women in Mathematics during their "relaxing" day. (A second group of five mathematics teachers from Washington, D.C., middle and high schools also visited the exhibit a few days later.) Many of these undergraduate mathematics majors ultimately teach mathematics at the secondary, undergraduate, or graduate level, so Amy decided to look for historical examples of significant mathematical ideas that might come up in a mathematics classroom. She consulted a few general works to complement personal brainstorming. [7; 11; 15] She had the Dibner's collection of approximately 10,000 rare books that cover all of the sciences to draw upon, and she placed the books she chose in three 5-foot by 2-foot glass cases in the library's reading room. Items in the exhibit were supported by the library's triangular book cushions and by plexiglass stands reused from earlier exhibits.

Given a general theme and these limitations of space and objects, she thought about what she wished to show. She wanted to use a variety of types of material (including early printed books, scholarly journals, paperback textbooks, and manuscript letters) by a variety of mathematicians (including as many women and as many English-language items as possible). Of course, she also hoped to provide maximum visual appeal. A treasure in the history of mathematics, the 1482 first printed edition of Euclid's *Elements of Geometry* (edited by Campanus) was displayed to illustrate the Fundamental Theorem of Arithmetic. Although the

Dibner owns the first edition of Diophantus' *Arithmetica* to incorporate Pierre de Fermat's famous marginal note into the printed text, Amy decided to represent Fermat's Last Theorem with a letter by Sophie Germain about her non-admittance to a scientific society to remind the students that women contributed to the attempts to solve this problem and to point out that these contributions did not necessarily lead to full participation in mathematical communities. Instead of showing a page of Latin text that would be mysterious to many American undergraduates, the figure from Gottfried Leibniz's 1684 *Acta Eruditorum* article, "Nova Methodus pro Maximis et Minimis," was chosen to represent Differentiation. Each book was measured while it was lying open before final decisions about arrangement were made.

Finding out about objects

Once you have a thesis or theme in mind, the next step is to learn about the objects you intend to exhibit. Begin with an examination of all parts of one of the objects, taking careful notes on what you observe. Consider not only what an object is, but any names or numbers associated with it. For example, if the object was produced commercially, does it have a model and/or serial number? This information could help pinpoint its date and place of origin. What do you know about the history of the artifact? Is there something about its previous owners or the way that it was used that makes it different from other things of the same type? Read any directions or documents that came with the object. As you acquire information, figure out a way that you can keep track of it, such as a card file or database.

There are many experts who can assist you with this process. Make friends with your librarians, colleagues, and, if there is one, the archivist at your institution. There may be instruction manuals as well as records explaining the purchase date and location and the reason for acquiring an object in your department. An archivist may have photographs documenting the use of your object or, more broadly, the history of activities in your department. Such visual images can add greatly to the appeal of an exhibit. College catalogues can provide further information about the history of your department's curriculum, as well as the names and educational attainments of faculty. Catalogues may also contain references to instruments owned by the mathematics department. Remember, too, that objects sometimes find their way into an archive or library.

You can use the bibliography at the end of this chapter to locate print and Web resources related to your objects. It may even be possible to contact specialists familiar with specific objects, such as slide rules or computers. This is one area where student help can be readily enlisted, but be sure to evaluate the veracity of any information found on the Web.

During this stage of the process, you may need the guidance of a conservator. [60; 61; 62; 63; 64; 65] Objects, especially those made from textiles or paper, are easily damaged by exposure to light or improper mounting. Book exhibits are generally very dimly lit for this reason. Of course, it is possible to use an image of a text rather than the text itself. You also may have the good fortune to have a duplicate of an object that you consciously use as a prop, knowing that it will probably suffer from being exhibited. For example, a piece of sheet music included in "Slates, Slide Rules" was such a duplicate. A conservator can make suggestions about how best to mount objects. In fact, an exhibition can offer an excuse to transform objects in very poor condition into something closer to their original state. This means that you will have to learn what objects should look like so a conservator can do his or her work. Figure 3 shows a model of an elliptic cone and hyperboloid of one sheet that came to the museum without strings, restrung for exhibition.

If you have the good fortune to hire a designer, he or she will need color photographs and measurements of each object, as well as an opportunity to see what you intend to exhibit. The designer also needs warnings about any objects requiring low light levels as well as copies of graphics. He or she will specify how images should be cropped and/or enlarged. Of course, the exhibit plan will also include the text that you plan to include about each object, which brings us to another topic.

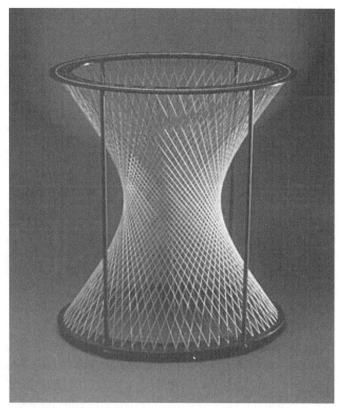

Figure 3. Geometric model manufactured in 1893 by Ludwig Brill of Darmstadt, Germany, restrung for exhibit. Gift of Wesleyan University. Negative number 2001-11865. Courtesy of the Smithsonian's National Museum of American History.

Writing labels

In a temporary exhibit, it is quite possible to simply tell your visitors orally what each object is. However, for the sake of your visitors who wander by a few minutes late, as an aid and supplement to what you say, and as a record of what you have done, it is preferable to refer to your ordering scheme and write up some of what you have learned into object labels. This is an especially important exercise if students are preparing the exhibit.

Do not assume that your audience has foreknowledge of your objects or even of mathematics. You will want to write text that is true but brief, clear but concise. Most viewers will not spend much time on reading. In addition, because viewers will be standing some distance away from the exhibit, you will need to choose a very large type size. The Smithsonian's Guidelines for Accessible Exhibition Design recommend a minimum type size of 24 points for a viewing distance of 3 inches, 48 points for 39 inches, 100 points for 78 inches, and 148 points for 118 inches. Choose a standard, readily legible font. Use a line length for text that facilitates reading; a maximum of 55 characters per line is a good rule to follow. You can divide your text into columns, although a maximum length of 100 words per label often works well. Do not set text in all capital letters, and avoid the use of script and italic type for essential information. If you expect to have a substantial number of visitors with disabilities, consider alternative forms of labels: large print versions, docents, Braille documents, audio versions. Place objects and labels where they can be seen by children, short people, and visitors in wheelchairs. Think about whether and how you wish to acknowledge those who have provided objects, graphics, and other assistance in preparing your show.

To write good labels, start with more information than you need, write, edit out some items, ask others to read your labels, and proofread again. What fundamental background information, historical context, and mathematical context do you need to convey? (One alternative when you have much more information to share than is manageable for an exhibit is to prepare handouts for those viewers who want to learn more.) Each label ought to be self-contained. At the same time, though, try to provide a narrative that draws viewers through the exhibit in an orderly fashion. You should select a title for your exhibit. You may want to provide a main label that introduces viewers to your thesis, and section labels for the different parts of the exhibit. Do not, on the other hand, turn this process into an obsession. Invest a level of effort appropriate to the scope and scale of your exhibit. Indeed, creating labels is not necessarily an expensive aspect of the exhibit: while the labels for "Slates, Slide Rules" were prepared professionally, the labels for "What Every Math Major Needs to Know" were printed on the back side of scratch tagboard (one example of a low-key label appears in Figure 4).

The Pythagorean Theorem
If triangle ACB is a right triangle, then $a^2 + b^2 = c^2$.

Nasir al-Din al-Tusi. *Kitab tahrir usul l-Uqlidus.* **Rome, 1594.**
Awareness of the special relationship between the sides and hypotenuse of a right triangle is one of the oldest ideas in mathematics, dating to at least 2000 B.C. The ancient Greeks required logical proofs of the mathematical concepts they knew. For example, Euclid made this theorem the climax of Book I of his *Elements of Geometry* (c. 300 B.C.); it often appeared with the classic windmill diagram you see here. Arab writers like Nasir al-Din (1201-1274) translated and commented on Greek works. This edition of Euclid's *Elements* was studied by European geometers including Girolamo Saccheri (1667-1733).

Figure 4. Exhibit label from "What Every Math Major Needs to Know."

Publicizing your exhibit

As you plan your exhibit, think about how it might attract visitors. Enticing objects arranged elegantly in a prominent location are a natural draw. So are related events, particularly events involving food (not, however, food that can be deposited on the objects). Time your exhibit to coincide with a meeting of the MAA, a colloquium on a related topic, or campus events such as Homecoming or Commencement. Send out fliers and electronic announcements. Tell the admissions office why every prospective student and student parent should see this glimpse into your institution's vibrant past. Think about school groups, especially at times of the year when teachers are especially inclined to take field trips (October, March, April and May are particularly good). Pass along information about the exhibit to institutional and community newspapers. Remember that if you want to attract visitors, you need to start this publicity before the exhibit is ready. This is yet another good reason to have a strong organizing scheme from the beginning.

Conclusion

In this chapter, we have provided a general overview of the process for identifying objects and organizing them into exhibits. We hope you can apply these suggestions to your own specific situation. Of course, looking at objects may well lead you to endeavors other than exhibits. You may find some apparatus that still is useful in the classroom. You might decide to produce an on-line catalog of your collections, as the University of Arizona's mathematics department did. [37] You might also find, like us and other authors of articles in this volume, that the objects you study lead you to publish research in the history of mathematics and mathematical instruments. In all these endeavors, we wish you the best!

Acknowledgments

The authors are indebted to the Smithsonian's Office of Imaging and Photographic Services for images. They also thank numerous NMAH staff, Ronald Brashear, formerly of Smithsonian Institution Libraries, and Murli Gupta of George Washington University for their assistance and recommendations. A preliminary version of this chapter was presented to the History of Mathematics Special Interest Group of the Mathematical Association of America in January 2004.

Annotated Bibliography

This bibliography is intended especially for those planning exhibits of mathematical instruments and books found in American colleges and universities. It includes some of the sources that would be of use in identifying, interpreting, and exhibiting objects that might be found on a campus in the United States. A few works that describe other kinds of objects are included as examples of how information can be presented. The focus of the bibliography is on planning the themes of an exhibit and identifying objects, not on conservation, design, or exhibit fabrication.

I. General works in history of mathematics, computing, and mathematical instruments

1. Amy Ackerberg-Hastings, Peggy A. Kidwell, and David L. Roberts, *Material to Learn: Tools of American Mathematics Teaching, 1800–2000*, The Johns Hopkins University Press, Baltimore, 2008.

2. Joe Albree, David C. Arney, and Frederick V. Rickey, *A Station Favorable to the Pursuits of Science: Primary Materials in the History of Mathematics at the United States Military Academy*, American Mathematical Society & London Mathematical Society, Providence, R.I., 2000. A catalog of one important collection of mathematics books, with much interesting material about mathematics teaching at West Point.

3. William Aspray, ed., *Computing before Computers*, Iowa State University Press, Ames, 1990. A collection of useful articles about pre-electronic computing devices.

4. James A. Bennett, *The Divided Circle: A History of Instruments for Astronomy, Navigation and Surveying*, Phaidon Christie's, Oxford, 1987. This book is not explicitly about mathematical instruments, but it touches on related topics.

5. Carl B. Boyer, *A History of Mathematics*, Uta C. Merzbach, rev., John Wiley & Sons, New York, 1991. A very useful general work.

6. Florian Cajori, *The Teaching and History of Mathematics in the United States*, Government Printing Office, Washington, D.C., 1890. A classic survey of mathematics teaching in the U.S. up to 1890.

7. William Dunham, *Journey Through Genius: The Great Theorems of Mathematics*, Penguin Books, New York, 1991. History of mathematics "highlights" that could be used as the basis for an exhibit.

8. Peter Duren, Richard A. Askey, and Uta C. Merzbach, eds., *A Century of Mathematics in America*, 3 vols., American Mathematical Society, Providence, R.I., 1988. Contains a variety of reference and source material.

9. Walther Dyck, *Katalog mathematischer und mathematisch-physikalischer Modelle, Apparate und Instruments*, C. Wolf & Sohn, Munich, 1892. The catalog of one of the first exhibits devoted exclusively to mathematical instruments.

10. Clark A. Elliott, compiler, *Biographical Index to American Science: The Seventeenth Century to 1920*, Greenwood Press, New York, 1990. Includes mathematicians.
11. John Fauvel and Jeremy Gray, eds., *The History of Mathematics: A Reader*, Macmillan Education in association with the Open University, Houndmills, Basingstoke, Hampshire, 1987. Source material with historical context.
12. Charles C. Gillispie, ed., *Dictionary of Scientific Biography*, 18 vols., Scribner, New York, 1970–1980. A multi-volume biographical encyclopedia that includes mathematicians.
13. E. M. Horsburgh, ed., *Handbook of the Napier Tercentenary Celebration or Modern Instruments and Methods of Calculation*, G. Bell and Sons and the Royal Society of Edinburgh, Edinburgh, 1914. This is an overview of instruments of calculation available in 1914, as well as historic objects exhibited at the celebration of the 300th anniversary of the publication of John Napier's tables of logarithms. In 1982, the book was republished with a new introduction by Michael R. Williams.
14. Louis C. Karpinski, with the cooperation of Washington Libraries of Walter F. Shenton, *Bibliography of Mathematical Works Printed in America through 1850*, The University of Michigan Press, H. Milford, Oxford University Press, Ann Arbor and Oxford, 1940. An annotated listing of early American mathematics books. There is a 1980 Arno reprint of this work.
15. Victor Katz, *A History of Mathematics: An Introduction*, HarperCollins, New York, 1993. General reference and textbook in the history of mathematics.
16. Peggy A. Kidwell and Paul E. Ceruzzi, *Landmarks in Digital Computing: A Smithsonian Pictorial History*, Smithsonian Institution Press, Washington, D.C., 1994. Brief accounts of a few important objects in the Smithsonian collections, with bibliography.
17. Morris Kline, *Mathematical Thought from Ancient to Modern Times*, Oxford University Press, New York, 1972. A survey work.
18. Henri Michel, *Scientific Instruments in Art and History*, R. E. W. Maddison and Francis R. Maddison, trans., The Viking Press, New York, 1967. This is one of several introductions to scientific instruments that makes an excellent coffee table book as well as contains useful information. The kinds of instruments shown in these books are relatively rare in American colleges. Other books of the same sort have been written by Nigel Hawkes and by Harriet Wynter and Anthony Turner.
19. Karen Hunger Parshall and David E. Rowe, *The Emergence of the American Mathematical Research Community 1876–1900: J. J. Sylvester, Felix Klein, and E. H. Moore*, American Mathematical Society & London Mathematical Society, Providence, R.I., 1994. A basic outline of the formation of the mathematical community in the U.S.
20. Deborah J. Warner, and Robert Bud, eds., *Garland Encyclopedia of Scientific Instruments*, Garland, Hamden, Conn., 1998. A collection of useful articles about various types of scientific and mathematical instruments, with bibliographies on different kinds of instruments.
21. Michael R. Williams, *A History of Computing Technology*, 2nd ed., IEEE Press, Los Alamitos, Cal., 1997. An outline of the history of computing devices to about 1970.

Related web sites:

22. Epact: Scientific Instruments of Medieval and Renaissance Europe, <www.mhs.ox.ac.uk/epact>. An online database of medieval and Renaissance scientific instruments made before 1600 A.D. that are in various European collections. It includes an encyclopedia that describes each kind of instrument shown.
23. Werner Girbardt und Werner H. Schmidt, Katalog der Rechentechnischen Sammlung des Instituts für Mathematik und Informatik Ernst-Moritz-Arndt-Universität Greifswald <www.uni-greifswald.de/~wwwmathe/RTS/>. Catalog of the mathematical instruments at this German mathematics institute. The instruments range from mechanical adding and calculating machines and slide rules to electronic calculators and personal computers.
24. Peggy A. Kidwell, curator, Slates, Slide Rules, and Software: Teaching Math in America, Smithsonian National Museum of American History, <americanhistory.si.edu/teachingmath>. The Web site of a Smithsonian exhibit on mathematics teaching devices.

25. David Kullman and Thomas Hern, eds., Ohio Masters of Mathematics, <www.bgsu.edu/departments/math/Ohio-section/bicen/>. Brief biographical accounts of mathematicians who had some association with the state of Ohio.

26. Museum of the History of Science, Oxford, <www.mhs.ox.ac.uk>. In addition to Epact (see above) this site has several admirable on-line exhibits and also an on-line collections database describing the collections at Oxford. There are also links to related Web sites.

27. John J. O'Connor and Edmund F. Robertson, The MacTutor History of Mathematics Archive, University of St. Andrews, <www-groups.dcs.st-andrews.ac.uk/~history/>. Contains numerous useful links.

II. Works on specific kinds of instruments

For information on specific makers, it often is most useful to consult their published catalogs. There are bibliographies in the *Encyclopedia of Scientific Instruments* listed above. [20] The bibliography of the Scientific Instrument Commission of the International Union of the History and Philosophy of Science is available electronically at <www.sic.iuhps.org/in_bibli.htm>.

a. Drawing instruments

28. Maya Hambly, *Drawing Instruments 1580-1980*, Sotheby Publications, London, 1988. A very useful book written for collectors.

29. William Ford Stanley, *Mathematical Drawing and Measuring Instruments*, E. & F. N. Spon, New York, 1888. An account by an eminent nineteenth-century English maker.

30. Several American scientific instrument dealers sold mathematical instruments. For information about them, one can consult: Physical Sciences Collection: Surveying and Geodesy, Smithsonian's National Museum of American History, <americanhistory2.si.edu/collections/surveying/>, a web site on Smithsonian surveying and geodesy collections.

31. There are also articles about various specific drawing instruments in *Rittenhouse: Journal of the American Scientific Instrument Enterprise*.

b. Geometric models

32. H. Martyn Cundy and A. P. Rollett, *Mathematical Models*, Clarendon Press, Oxford, 1961. This is not an historical source, but gives a useful description of several kinds of models.

33. Gerd Fischer, ed., *Mathematische Modelle aus den Sammlungen von Universitäten und Museen*, 2 vols., Vieweg&Sohr, Braunschweig, 1986. The first volume of this work shows photographs of models (mainly late nineteenth and early twentieth century German models by Brill and Schilling); the second provides mathematical interpretation. There is some introductory material on how the models were made.

34. History of Mathematics at the United States Military Academy, <www.dean.usma.edu/math/about/history/>. Includes images of nineteenth-century French models for descriptive geometry designed by Theodor Olivier.

35. Le Raccolte Museali Italiane di Modelli per lo Studio delle Matematiche Superiori, <www.dma.unina.it/~nicla.palladina/catalogo/>. An Italian on-line museum of geometric models (many of these are nineteenth-century German models), with photographs.

36. Peggy A. Kidwell, American Mathematics Viewed Objectively—The Case of Geometric Models, in *Vita Mathematica: Historical Research and Integration with Teaching*, Ronald Calinger, ed., Mathematical Association of America, Washington, D.C., 1996, pp. 197–208.

37. Mathematical Teaching Tools in the Department of Mathematics, The University of Arizona, <www.math.arizona.edu/~models/>. An on-line catalog of the models, calculating machines and other teaching devices at the University of Arizona.

38. Karen Parshall and David E. Rowe, Embedded in the Culture: Mathematics at the World's Columbian Exposition of 1893, *The Mathematical Intelligencer* 15, no. 2 (1993) 40–45.

39. David L. Roberts, Albert Harry Wheeler (1873–1950): A Case Study in the Stratification of American Mathematical Activity, *Historia Mathematica* 23 (1996) 269–287.
40. Amy Shell-Gellasch, The Olivier String Models at West Point, *Rittenhouse*, 17 (2003) 71–84.
41. University of Toronto Museum of Scientific Instruments, <www.chass.utoronto.ca/utmusi/>. This Web site has a small section on geometric models, many from the collection of H. S. M. Coxeter.
42. Magnus J. Wenninger, *Polyhedron Models*, The University Press, Cambridge, 1971. Instructions for model makers.

c. Slide rules

43. Florian Cajori, *A History of the Logarithmic Slide Rule and Allied Instruments*, Astragal Press, Mendham, N.J., 1994. This is a reprint of the original edition of 1910. Cajori's classic work was also published in several editions of W. W. Rouse Ball's *String Figures and Other Monographs*.
44. Peter M. Hopp, *Slide Rules: Their History, Models and Makers*, Astragal Press, Mendham, N.J., 1999. This book was written primarily with collectors in mind.
45. Dieter von Jezierski, *Slide Rules: A Journey Through Three Centuries*, Rodger Shepherd, trans., Astragal Press, Mendham, N.J., 2000. This book, again published for collectors, is a translation of a text published in German in 1977. The author was the marketing manager of A. W. Faber-Castell for 40 years.
46. *Journal of the Oughtred Society*. A publication aimed primarily at slide rule collectors.

d. Planimeters and other integrators

47. Joachim Fischer, Instrumente zur mechanischen Integration: Ein Zwischenbericht, in *Brückenschläge: 25 Jahre Lehrstuhl für Geschichte der exakten Wissenschaften und der Technik an der Technischen Universität Berlin*, Hans-Werner Schütt and Burghard Weiss, eds., Engel, Berlin, 1995, pp. 111–156.
48. H. de Morin, *Les Appareils d'Intégration.* Gauthier-Villars, Paris, 1913.

e. Calculating machines and other digital devices

49. David J. Brydan, *Napier's Bones: A History and Instruction Manual*, Harriet Wynter Ltd, London, n.d. A brief account of an aid to computation used in the seventeenth and eighteenth centuries.
50. Ernst Martin, *Die Rechenmaschinen und ihre Entwicklungsgeschichte*, Johannes Meyer, Pappenheim, 1925. An account of the calculating machines available to that time. An English translation was published in 1992.

f. Other Teaching Apparatus, Puzzles and Educational Games

51. Charnel Anderson, *Technology in American Education 1650–1900*, Washington, D.C.: Government Printing Office, 1960.
52. Karen Hewitt and Louise Roomet. *Educational Toys in America: 1800 to the Present*, Burlington, Vt.: University of Vermont, 1979. This is an exhibit catalog.
53. Peggy A. Kidwell, An Erasable Surface as Instrument and Product: The Blackboard Enters the American Classroom—1800–1915, *Rittenhouse* 17 (2003) 85–98.

g. Electronic computers and calculators

54. Janet Abbate, *Inventing the Internet*, MIT Press, Cambridge and London, 1999. This book would be of use if one wished to consider the influence of computers on mathematical communication.
55. Guy Ball and Bruce Flamm, *The Complete Collector's Guide to Pocket Calculators*, Wilson/Barnett Publishing, Tustin, Cal., 1997.
56. Larry Cuban, *Oversold and Underused: Computers in the Classroom*, Harvard University Press, Cambridge, 2001. This book does not have any particular discussion of mathematics teaching.

57. Larry Cuban, *Teachers and Machines: The Classroom Use of Technology Since 1920*, Teachers College, New York and London, 1986. This volume talks about the use of radio, television and the computer in teaching, but not necessarily mathematics teaching.
58. Thomas F. Haddock, *A Collector's Guide to Personal Computers and Pocket Calculators (1956–1991)*, Books Americana, Florence, Ala., 1993.
59. The journal *IEEE Annals of the History of Computing* includes articles about a wide array of electronic and pre-electronic computing devices.

III. Conservation of objects

Here you would do well to consult with librarians (especially those in Special Collections) and any museum staff who might be associated with your institution. Other sources include:

60. Canadian Conservation Institute, *CCI Notes.* Short notes on the conservation of specific types of objects.
61. American Institute for Conservation, *Directory of American Institute for Conservation of Historic and Artistic Works*. This is an annual publication that lists qualified conservators. For further information, see <aic.stanford.edu>.
62. Per E. Guldbeck, *The Care of Historical Collections: A Conservation Handbook for the Nonspecialist*, American Association for State and Local History, Nashville, 1972. Described by Smithsonian conservators as an "oldie but goodie" that might be available through interlibrary loan.
63. Jane S. Long and Richard W. Long, *Caring for Your Family Treasures*, Heritage Preservation, Washington, D.C., 2000. For ordering information and related publications see the Heritage Preservation web site at <www.heritagepreservation.org/>.
64. National Park Service, *Conserve O Grams*, U.S. Government Printing Office. Notes on specific topics.
65. National Park Service, *Exhibit Conservation Guidelines.* A general and helpful compact disc.

IV. Exhibit design & material culture

66. Asa Briggs, *Victorian Things*, University of Chicago Press, Chicago, 1988. A general historical discussion of how manufactured objects became common in the nineteenth century.
67. Creating a Classroom Museum, Smithsonian Center for Education and Museum Studies, <smithsonianeducation.org/educators/lesson_plans/collect/crecla/crecla0a.htm>. This web site describes steps elementary school students might take in creating a classroom exhibit.
68. Steven Lubar and Kathleen M. Kendrick, *Legacies: Collecting America's History at the Smithsonian*, Smithsonian Institution Press, Washington, D.C., 2001. Coffee table book that distinguishes between artifacts and relics, while suggesting how objects can be informative.
69. Beverly Serrell, *Exhibit Labels: An Interpretive Approach,* Alta Mira Press, Walnut Creek, 1996.
70. Lothar P. Witteborg, *Good Show!: A Practical Guide for Temporary Exhibitions*, Andrea P. Stevens, ed., Steven D. Schindler, illus., Smithsonian Institution Traveling Exhibition Service, Washington, D.C., 1991. This is a nuts and bolts guide on planning, fabricating, installing, securing and publicizing a temporary exhibit. It assumes that the exhibit itself is coming from elsewhere. Includes bibliography.

About the Authors

Lieutenant Colonel William Acheson graduated from the United States Military Academy (USMA) at West Point in 1990 with a degree in Systems Engineering. After completing his master's from the Georgia Institute of Technology in 2000, he returned to USMA to teach in the Department of Mathematical Sciences from 2000 to 2002. At the time of publication LTC Acheson is assigned to a Military Transition Team advising the Iraqi Army as part of Operation Iraqi Freedom.

Amy Ackerberg-Hastings teaches history online for University of Maryland University College. Her research interests emphasize the history of nineteenth-century American mathematics education and the history of women in science and mathematics. Previous publications include "Protractors in the Classroom: An Historical Perspective," in *From Calculus to Computers: Using the Last 200 Years of Mathematical History in the Classroom*, eds. Richard Jardine and Amy Shell-Gellasch, (Mathematical Association of America Notes No. 68, Washington, DC, 2005), 217-228; and "Analysis and Synthesis in John Playfair's Elements of Geometry," *British Journal for the History of Science* 35 (2002): 43–72.

Len Berggren earned his doctorate in mathematics at the University of Washington in 1966 and spent about 2 1/2 years in the early and mid-70s at what was then the Department of History of Science and Medicine at Yale University, learning how to work in the history of ancient and medieval mathematics. Since 1966 he has been a member of the Department of Mathematics at Simon Fraser University, where he is Professor. He is a member of, among other organizations, the MAA and the North American Sundial Society

His general research area is the history of mathematical sciences, such as geography, cartography, astronomy, and time keeping—all areas where mathematics has interacted in important ways with the wider culture. He is completing a project of translating and publishing the collected works of the tenth-century Islamic mathematician, Abu Sahl al-Kuhi. Selected books include *Episodes in the Mathematics of Medieval Islam, Pi: A Source Book* (with J. and P. Borwein) and *Ptolemy's Geography: An Annotated Translation of the Theoretical Chapters* (with Alexander Jones).

James Evans is co-director of the Program in Science, Technology, and Society at the University of Puget Sound. He is the author of *The History and Practice of Ancient Astronomy* (Oxford U. P., 1998) and (with J. Lennart Berggren) of *Geminos's 'Introduction to the Phenomena': A Translation and Study of a Hellenistic Survey of Astronomy* (Princeton U.P., 2006), as well as the editor (with Alan Thorndike) of *Quantum Mechanics at the Crossroads: New Perspectives from History, Philosophy and Physics* (Springer, 2006).

Robert Foote received his Ph.D. from the University of Michigan, and has been on the faculty at Wabash College since 1989. His research interests are in differential geometry. Many years ago an engineering student showed him a planimeter, and he has been studying the mathematics of planimeters and collecting them ever since.

Peggy Aldrich Kidwell is Curator of Mathematics at the Smithsonian's National Museum of American History. She is an historian of science by training, and cares for a collection of objects that range from astrolabes to geometric models to geoboards.

Katherine Inouye Lau is a graduate of Brown University where she majored in Visual Arts. She is passionate about art, mathematics, writing, pedagogy and the areas where disciplines intersect. After taking Dr. Plofker's course, she was a teaching assistant for Professor Thomas Banchoff's course on geometry in four and higher dimensions, "The Mathematical Way of Thinking." Katherine is currently a graphic designer in California.

Major Brian Lunday was an Instructor and then Assistant Professor at the Department of Mathematical Sciences at the United States Military Academy from 2001–2004. He will begin studies for a Ph.D. in Industrial and Systems Engineering at the Virginia Polytechnic Institute beginning in 2007, to be followed by additional service as an Assistant Professor at USMA.

Hugh McCague is a mathematician and art historian. His research and publications have concentrated on the application of mathematics in ancient and medieval art and architecture. He has taught applications of computing in the arts at York University, Toronto, Canada.

Joanne Peeples received her B.S. and M.S. from Wichita State University, and then traveled to Germany to study at the University of Frankfurt. She completed her Ph.D. at New Mexico State University. In the summers of 1998 and 1999 she attended the Institute for the History of Mathematics and Its Use in Teaching (IHMT). Joanne currently teaches at El Paso Community College, where all of her classes complete math history projects. She is active in MAA—both at the national level and the section level, as well as in AMATYC (where she is a consultant for the current ACCCESS cohort) and at the local level is president-elect for NMMATYC

Kim Plofker is a historian of exact sciences specializing in the mathematics and astronomy of South and West Asia, as well as early modern Latin mathematics. Among her recent publications is the chapter "India" in *Mathematics of Egypt, Mesopotamia, China, India and Islam: A Source Book*, edited by Victor J. Katz, appearing in summer 2007 from Princeton University Press. She has been involved in numerous projects on using the history of mathematics in mathematics pedagogy, including the January 2003 MAA Short Course "Mathematics in the Ancient World".

This article is an expanded version of Ms. Lau's final project for a course called "Calculus and Its History" that Dr. Plofker taught in the Mathematics and History of Mathematics Departments at Brown University in 2002.

V. Frederick Rickey, a logician turned historian, became Professor of Mathematics at the United States Military Academy, West Point, NY in the summer of 1998. After earning three degrees from the University of Notre Dame (Ph.D. 1968) he went to Bowling Green State University where he rose through the professorial ranks to the rank of Distinguished Teaching Professor Emeritus. He has broad interests in the history of mathematics and is especially interested in the development of the calculus.

He has been on leave five times, most recently in Washington D. C. where he was Visiting Mathematician at the MAA Headquarters. While there he was involved in the founding of *Math Horizons,* a magazine for mathematics undergraduates; became the first editor of electronic services for the MAA and built its first gopher and web pages; and wrote a successful NSF grant for an Institute for the History of Mathematics and Its Use in Teaching.

He loves teaching and enjoys giving lectures to mathematicians about the history of their field. He received the first award from the Ohio Section for Distinguished College or University Teaching of Mathematics, and one of the first MAA National Awards for teaching.

Ed Sandifer took his PhD in Mathematics, studying algebra under John Fogarty at UMass Amherst. Victor Katz and Fred Rickey diverted his interests at their Institute for the History of Mathematics and Its Uses in Teaching (IHMT) over several summers in Washington, DC. Since then he has concentrated on Euler, Spanish Colonial mathematics, and now on planimeters and integrating devices. He is an enthusiastic long distance runner and has run the Boston Marathon more than 30 times.

Amy Shell-Gellasch is currently a faculty fellow at Pacific Lutheran University in Tacoma, WA. She is the Programs Chair for the History of Mathematics Special Interest Group of the MAA as well as Chairing the MAA Committee on SIGMAA's. She received her bachelor's degree from the University of Michigan in 1989, her master's degree from Oakland University in Rochester, Michigan in 1995, and her doctor of arts degree from the University of Illinois at Chicago in 2000. Her dissertation was a biographical piece on mathematician Mina Rees. Most recently, she conducted research with V. Fredrick Rickey on the history of the Department of Mathematical Sciences at the United States Military Academy, where she was an Assistant Professor.

Daina Taimina was born and received all her formal education in Riga, Latvia. In 1977 she started to teach at the University of Latvia and continued there for more than 20 years. Her PhD thesis was in mathematics and theoretical computer science under the supervision of Rusins Freivalds. Later she got more involved with geometry, history of mathematics, and mathematics education. Since 1997 Daina has worked at Cornell University as a Visiting Associate Professor and a senior research associate. She is co-author (with David W. Henderson) of *Experiencing Geometry: Euclidean and non-Euclidean with strands of history*, 3rd edition, Prentice Hall, 2004.

David E. Zitarelli has been at Temple University since obtaining his Ph.D. in 1970 in algebraic semigroups under Mario Petrich at Penn State. In 1973 Kenneth May invited him to attend a conference at Texas Tech on American mathematics, which began his conversion to the history of mathematics. He served as abstracts editor of *Historia Mathematica* from 1988 to 2000, organized six special sessions at AMS-MAA meetings, and taught several Chautauqua courses on the history of mathematics. 2001 was a wonderful odyssey, as Z published a paper on towering figures in American mathematics and a book on the history of an MAA section, a video of his lecture on the genesis of the Moore Method was produced and distributed, and he won two teaching awards. He also organized the Philadelphia Area Seminar on the History of Mathematics that year (with Thomas Bartlow) that continues to hold monthly meetings. The following year he was a Buckingham Scholar-in-Residence at Miami University in Ohio. In 2003 he was elected the first chair of HOMSIGMAA, the MAA special interest group on the history of mathematics. His paper on the travails of Joseph B. Reynolds and Lehigh University to come to grips with a research agenda will appear in *Historia Mathematica* in 2007. A paper on Moore School member Anna Mullikin (with PASHoM co-founder Bartlow) is in the works. Z presented his preliminary findings at a conference in Lhasa, Tibet, in July 2007. In his pre-history period he wrote a series of textbooks on finite mathematics and calculus (with Raymond Coughlin), and his linear algebra lab manual (with David Hill) warranted an invitation to deliver a one-hour address at the 1995 MAA annual meeting.